The Electrochemical Society

Analytical Techniques for Semiconductor Materials and Process Characterization 5

(ALTECH 2007)

ECS Transactions Volume 10 No.1

September 13-14, 2007
Munich, Germany

Printed from e-media with permission by:

Curran Associates, Inc.
57 Morehouse Lane
Red Hook, NY 12571
www.proceedings.com

ISBN: 978-1-60423-825-9

Some format issues inherent in the e-media version may also appear in this print version.

Copyright 2007 by The Electrochemical Society, Inc.
All rights reserved.

This book has been registered with Copyright Clearance Center, Inc.
For further information, please contact the Copyright Clearance Center,
Salem, Massachusetts.

Published by:

The Electrochemical Society, Inc.
65 South Main Street
Pennington, New Jersey 08534-2839, USA

Telephone 609.737.1902
Fax 609.737.2743
e-mail: ecs@electrochem.org
Web: www.electrochem.org

Printed in the United States of America.

Analytical Techniques for Semiconductor Materials and Process Characterization 5 (ALTECH 2007)

Editors:

B. O. Kolbesen
J. W. Goethe Universität
Frankfurt, Germany

C. L. Claeys
IMEC
Leuven, Belgium

L. Fabry
Wacker Chemie
Burghausen, Germany

Sponsoring Division:

 Electronics and Photonics

Published by
The Electrochemical Society

65 South Main Street, Building D
Pennington, NJ 08534-2839, USA

tel   609 737 1902
fax   609 737 2743
www.electrochem.org

ecstransactions ™

Vol. 10 No. 1

PREFACE

The *ALTECH 07* symposium *Analytical Techniques for Semiconductor Materials and Process Characterization V*, was held as part of the Solid State Device Research Conference (ESSDERC) 2007 in Munich, Germany on September 13-14, 2007.

Since 1989, when the first *ALTECH* symposium was held in conjunction with the 19th *ESSDERC* in Berlin, diagnostic capabilities have advanced extensively. Without smart and affordable diagnostics, the semiconductor industry could not have grown to its market value of more than $240 billion in 2006. This ALTECH issue of *ECS Transactions* contains 14 papers, focusing on testing engineered silicon materials, defects and contamination characterization, nanotechnology, sophisticated statistical treatment of data near the limit of detection, thin film metrology, electron microscopy, and in-line electric testing.

The symposium was organized into five sessions:

- Etching techniques for defect delineation
- Metrology
- EU Project: Analytical Network for Nanotech (ANNA)
- Techniques for characterization of defects and contamination
- Electrical and scanning probe microscopy

These activities represent activities in various industrial and academic research institutes in Europe. Fortunately, the relative contribution of European businesses in the world market diagnostic equipment is much stronger than its market share of the world semiconductor market. Hence, we present an optimistic perspective of further developments in Europe.

The *Analytical and Diagnostic Techniques for Semiconductor Materials, Devices, and Processes / VII*, sponsored by the Electronics and Photonics Division was part of the Fall 2007 Electrochemical Society Meeting in Washington, DC on October 7-12, 2007. It is the seventh meeting in a series of symposia presenting the international activities in this field since 1988.

Twenty four of the 42 papers presented at this symposium, covering a wide array of topics, were invited papers. Some gave overviews in their field, while others presented recently developed measurements techniques. The techniques spanned electrical, optical, electron beam, x-ray, ultrasonic, contactless, and atom probe techniques; additionally:

- Defect characterization techniques include photoluminescence, infrared (IR) absorption, IR laser scattering tomography, microwave-photoconductance decay, surface photovoltage, contact potential difference, deep-level transient spectroscopy, current-voltage, atomic force microscopy and conductance AFM probe microscopy, ultrasonics for vacancy measurements, and chemical etching
- High-resolution transmission electron microscopy and atom probe measurements allow visualization of individual atoms and their distribution in the sample
- Electron holography allows the internal potential distribution in semiconductor devices to be visualized

- Recent advances in x-ray metrology, *e.g.*, x-ray fluorescence, x-ray reflection, and synchrotron radiation, were highlighted in several papers

We thank the Electrochemical Society staff members John Lewis, Stephanie Plassa, Paul Urso, and Beth Anne Stuebe for their support and advice, the reviewers of the papers, and the speakers who contributed their papers to these Proceedings.

ALTECH 2007:
Bernd O. Kolbesen
Cor L. Claeys
Laszlo Fabry

Symposium E3 Diagnostics:
Dieter K. Schroder
Laszlo Fabry
Richard S. Hockett
Alain C. Diebold
Hirofumi Shimizu

October 2007

ECS Transactions, Volume 10, Issue 1
Analytical Techniques for Semiconductor Materials and Process Characterization 5
(ALTECH 2007)

Table of Contents

Preface

Chapter 1
Etching Techniques for Defect Delineation

A Review of Different and Promising Defect Etching Techniques: from Si to Ge *   3
   A. Abbadie, J. Hartmann and F. Brunier

Organic Peracid Etches: A New Class of Chromium Free Etch Solutions for the   21
Delineation of Defects in Different Semiconducting Materials
   D. Possner, B. O. Kolbesen, H. Cerva and V. Kluppel

Chapter 2
European Project ANNA (Analytical Network for Nanotech)

Metrology, Analysis and Characterization in Micro- and Nanotechnologies –   35
A European Challenge *
   L. Pfitzner, A. Nutsch, R. Oechsner, M. Pfeffer, E. Don, C. Wyon and
   M. Hurlebaus

Wafer Contamination Analysis, Speciation and Reference-free Nanolayer   51
Characterization using Synchrotron Radiation based X-ray Spectrometry *
   B. Beckhoff, R. Fliegauf, M. Kolbe, M. Müller, B. Pollakowski, J. Weser and G.
   Ulm

Structural and Analytical Characterization by Scanning Transmission Electron   57
Microscopy of Silicon-based Nanostructures
   A. Armigliato, R. Balboni and A. Parisini

Thin Film Analysis and Model Interface Characterization Studies of Relevance to   65
Microelectronics
   I. Dontas, V. Papaefthimiou, S. Kennou and S. Ladas

Surface Microdefects Control during Chemical Mechanical Polishing of Silicon   75
Wafers: an Example of in line Manufacturing Process Control
   G. Borionetti, A. Corradi, N. Mainardi, A. M. Rinaldi and K. Takami

Molybdenum Contamination in Indium and Boron Implantation Processes   85
   D. Codegoni, M. Polignano, V. Soncini and C. Bresolin

Chapter 3
Techniques for Characterization of Defects and Contamination

Application of Selected Electron Microscopy Methods to Materials Analysis   97
Problems *
   A. Rucki and H. Cerva

Characterization of Nickel-related Defects in Thin SOI Substrates after Thermal   109
Treatment
   I. Rink and C. Emons

Detailed Photocurrent Analysis of Iron Contaminated Boron Doped Silicon by   117
Comparison of Simulation and Measurement
   M. Rommel, A. Bauer and H. Ryssel

Chapter 4
Electrical and Scanning Probe Microscopy Techniques

Structural and Electrical Characterization of Dielectrics, Carbon Nanotubes and   129
Nanoelectronic Devices by Means of Scanning Probe Microscopy *
   U. Schwalke

Methods for the Controlled Reduction of Carrier Lifetime in Power Devices *   141
   F. Hille, F. Niedernostheide and H. Schulze

In-Line Characterization of Heterojunction Bipolar Transistor Base Layers by   151
High-Resolution X-Ray Diffraction
   N. Nguyen, R. Loo, A. Hikavyy, B. Van Daele, P. Ryan, M. Wormington and J.
   Hopkins

Author Index   161

* = *invited paper*

Facts about ECS

The Electrochemical Society (ECS) is an international, nonprofit, scientific, educational organization founded for the advancement of the theory and practice of electrochemistry, electrothermics, electronics, and allied subjects. The Society was founded in Philadelphia in 1902 and incorporated in 1930. There are currently over 7,000 scientists and engineers from more than 70 countries who hold individual membership; the Society is also supported by more than 100 corporations through Corporate Memberships.

The technical activities of the Society are carried on by Divisions. Sections of the Society have been organized in a number of cities and regions. Major international meetings of the Society are held in the spring and fall of each year. At these meetings, the Divisions and Groups hold general sessions and sponsor symposia on specialized subjects.

The   publications program   the following.

Journal of The Electrochemical Society — JES is the peer-reviewed leader in the field of electrochemical and solid-state science and technology. Articles are posted online as soon as they become available for publication. This archival journal is also available in a paper edition, published monthly following electronic publication.

Electrochemical and Solid-State Letters — ESL is the
journal   areas   Articles are posted online   soon
                  publication.   -reviewed,   journal   in   paper
        published monthly following electronic publication. It is a joint publication of ECS and the IEEE Electron Devices Society.

Interface — *Interface* is ECS's quarterly news magazine.   provides a forum for the lively exchange of ideas and news   ECS and the international scientific community at large. Published online (with free access to all) and in paper,   special features on the   of electrochemical and solid-state science and technology The paper edition is automatically sent to all

Meeting Abstracts (formerly Extended Abstracts) — Abstracts of the technical papers presented at the spring and fall meetings of the Society are published on CD-ROM.

ECS Transactions — This online database provides access to full-text articles presented at ECS and ECS-sponsored meetings. Content is available through individual articles, or as collections of articles representing entire symposia.

Monograph Volumes — The Society sponsors the publication of hardbound monograph volumes, which provide authoritative accounts of specific topics in electrochemistry, solid-state science, and related disciplines.

For more information on these and other Society activities, visit the ECS website:

www.electrochem.org

CHAPTER 1

ETCHING TECHNIQUES FOR DEFECT DELINEATION

A review of different and promising defect etching techniques : from Si to Ge

A. Abbadie [(1)], J.M. Hartmann [(2)], and F. Brunier [(1)]

(1) SOITEC, Parc Technologique des Fontaines, 38190 Bernin, France;
(2) CEA-LETI, Minatec, 17, Rue des Martyrs 38054 Grenoble, France

A review of defect selective etchings used for process monitoring of Si, SiGe and Ge is presented. Cr-based and Cr-free solutions as well a gaseous HCl etch technique are described. Some applications of these techniques on Si and SiGe relaxed substrates as well their etching mechanisms are highlighted and discussed.

Introduction

The transfer of column IV layers on Si has recently been demonstrated using the Smart - Cut[TM] technology (1-4). It indeed allows thickness, Ge composition and crystal orientation variations and has successfully been used for the fabrication of highly competitive substrates for CMOS applications: Silicon On Insulator (SOI), HOT (Hybrid Orientation Transfer) SOI, strained and eXtra-strained Silicon On Insulator (sSOI and XsSOI), Germanium On Insulator (GeOI) etc (1-4). It is now extended to III-V materials for optoelectronic applications. The assessment of the crystalline quality of these materials requires the development of sensitive and reliable characterization methods. Selective chemical etching, being a fast and simple method with a large area sampling, is conventionally used for defects decoration. However, hexavalent chromium, which is present in the widely used solutions such as Secco or Schimmel, is very toxic. Cr-free new solutions have then to be developed for defects revelation in SiGe.

We have reviewed in this paper the different etching techniques (chemical wet-etches and gaseous HCl etch) that can be used for crystal quality monitoring (from the starting material (Si, SiGe or Ge) to the final product (i.e. column-IV layers on insulator)). We have notably focused on the SiGe virtual substrates that are used as templates for the fabrication of sSOI and XsSOI wafers. The analysis of their crystalline quality is primordial as it will directly impact the quality of the sSi layers grown on top and the quality of the final product (i.e. sSOI or XsSOI). In this study, Leti SiGe samples with different Ge concentrations (from 20% up to pure Ge) were chosen because of their known and well-calibrated defect densities, allowing a reliable comparison between the different techniques (5). Several techniques such as Electron Beam Induced Current (EBIC) or cathodoluminescence can be used to evaluate dislocations densities in relaxed SiGe (6). We have focused in this paper on wet-chemical solutions, in particular, diluted Schimmel ((CrO_3-H_2O)/HF), diluted Secco (($K_2Cr_2O_7$-H_2O)/HF) and a homemade Cr-free solution. Recently, new classes of Cr-free, more ecology-friendly etch solutions have indeed been developed, with good performances (i.e. an etch rate which is controlled, a spatial etching uniformity, a sensitivity to defects similar to the ones of Schimmel or Secco etc) (7-8). Finally, we have focused on crystalline defects revelation in SOI structures.

The starting samples

We have used for this study (i) high quality SiGe virtual substrates with Ge concentrations equal either to 20%, 30%, 40% or 50% (9) and (ii) 2.5 μm thick pure Ge layers grown using a low-temperature / high temperature approach on Si(001) followed by some thermal cycling (10). The nearly 1 μm thick constant composition SiGe layers sitting on top of our virtual substrates are nearly fully relaxed (macroscopic degrees of strain relaxation in-between 95% and 100%) and characterized by Threading Dislocations Densities around 10^5 cm^{-2}, typically. The Ge thick layers are in a slightly tensile strain configuration, with a TDD of the order of 10^7 cm^{-2} (11). Conventional SOI substrates samples obtained thanks to the Smart-Cut™ Technology were used for defect etching studies.

The etching solutions : from Si to Ge

Whatever the material, Si, SiGe or Ge, the etching process requires the presence of an oxidizing agent (characterized in general by its standard oxidation potential) and a fluoride component which aims at dissolving the oxidized material. The main oxidizing agents used are nitric acid (HNO_3), peroxide hydrogen (H_2O_2), and chromate in the form of hexavalent chromium (CrO_3, $K_2Cr_2O_7$). Some solutions require as well the presence of a dissolving agent (water, organic acid or acetic acid in the most usual cases) in order to dilute the solution and provide a control on the etching rates. The ratio between the different components will determine the rate-limiting step of the reaction and so the global mechanism of the reaction. The etching process can generally be separated in reaction controlled (general case) and diffusion controlled etchings (influenced by small changes in temperature, oxidants' nature and concentration etc..). Whatever the type of etchants used, defect selective etching requires a modification of the surface potential between the perfect crystalline material and defects.

Chromium-based etching solutions, such as Sirtl, Secco, Wright or Schimmel have become routine techniques, widely used for defects delineation in crystalline materials, including SiGe (12-15). The etching mechanism for Cr^{6+} - containing etching solutions is not well-known. A strong adsorption of chromates on the Si surface via silanol groups is expected, resulting in the formation of $Si-O-Cr(O_2)-O-Cr(O_2)-OH$ (16). The Cr-O bond strength (110 kcal.mol.$^{-1}$) is indeed among the highest bond strengths reported for chromium. Meanwhile, the Si-O bond strength (191 kcal.mol.$^{-1}$) is among the highest bond strengths for Si. For comparison, the Ge-O bond strength is of the order of 157 kcal.mol.$^{-1}$. Such bond strengths combined with the high standard oxidation potential of the Cr^{6+}/Cr couple could explain in part the high reactivity of chromium with the Si surface, resulting in quite high etch rates (>μm/min) compared to Ge surfaces (see figure 2 below). Such high etch rates are one of the limitations of the use of chromium, besides its high toxicity (Cr^{6+} is already forbidden in Asia).

The Dash solution ($HF/HNO_3/CH_3COOH$) was one of the first Cr-free etching solutions that enabled defects revelation in Si (17). Robbins and Schwartz (18) studied the mechanism of such solutions and attempted to explain the role of each component. Kashiwagi et al in 1996 (19) investigated defect delineation in $HF/HNO_3/CH_3COOH$ solutions with different concentrations. They notably studied the ratio of the defect etch rate to the bulk Si etch rate, the defects' shape etc. Recently, Steinert et al (20) and Acker et al (21) have studied the decomposition products of HNO_3 in order to explain the Si oxidation. They have concluded to a more complex reaction, in which HNO_3 is progressively reduced into a N(II) specie. The resulting and reactive specie NO^+ should be responsible for the Si oxidation, through the oxidative attack of Si-H and Si-Si bonds (equations [1] and [2]). The SiO_2 dissolution step is then realized with HF (22). It is also the fast step of the reaction (equation [3]).

$$NO + 4\,H_2O \leftrightarrow HNO_3 + 3\,H^+ + 3\,e^- \quad E^o\,(HNO_3/NO) = 0.96 \qquad [1]$$
$$3\,Si + 4\,HNO_3 \leftrightarrow 3\,SiO_2 + 4\,NO + 2\,H_2O \qquad [2]$$

$$SiO_2 + 6\,HF \leftrightarrow 2\,H^+ + SiF_6^{2-} + 2\,H_2O \qquad [3]$$

The etching in HF/HNO$_3$ containing solutions is an autocatalytic reaction. The proportion between HF volumes and HNO$_3$ volumes have also to be accurately defined, since the ratio between the different components might influence the overall mechanism of the reaction and especially the etch rate of the solution. For example, in a high hydrofluoric acid domain, HNO$_3$ is the kinetically important reagent. Its diffusion rate toward the Si surface at lower temperatures will limit the overall etching rate of the reaction. In a high nitric acid domain, HF plays an important role by controlling the oxide removal process and so the bulk etch rate.

Globally, a chemical etching is, as any chemical dissolution, an oxido-reduction process. Carriers are transferred from the substrate to the chemicals (oxidation step). New compounds form which are then dissolved. Such a reaction is close to the one of anodic dissolution. In the latter case, the carriers are supplied by a controlled external power source. The differential etch rate during anodic etching depends on the current density. At small current densities, defects etch much faster than perfect silicon, delineating the defects very clearly. Preferential etching can therefore be understood in terms of current flow. The notion of electronically active defects then becomes meaningful. More carriers will recombine on electronically active defects than in defect-free regions and the current will be locally reduced. The Electron Beam Induced Current (EBIC) method is a non destructive technique used for defects imaging (23-24).

For the sake of completeness, it should be mentioned that others Cr-free etches solutions have been developed in the nineties by MEMC (25-26). Such solutions, named "MEMC etches" have the particularity to be four-component solutions, in which both acetic acid and water are employed as diluents. The presence of copper-nitrate is very useful for defects differentiation in p- and n-type Si(100) and Si(111) substrates. Such solutions present numerous advantages such as the lack of any harmful metallic species, etch rates and behaviour which are similar to the ones of Sirtl or Wright etches, good pits definition. One of its main drawbacks is that it does not etch new engineered substrates such as "silicon on-insulator" wafers. As we will detail in the last part of the article, others solutions have then to be designed to overcome the non-etching behavior observed with the standard Cr-free etches initially described more than 10 years ago.

Finally another etching technique has recently appeared, which is based on the use of gaseous HCl in an epitaxy reactor (27). HCl can indeed be used to etch Si, SiGe and Ge, as shown in figure 1.

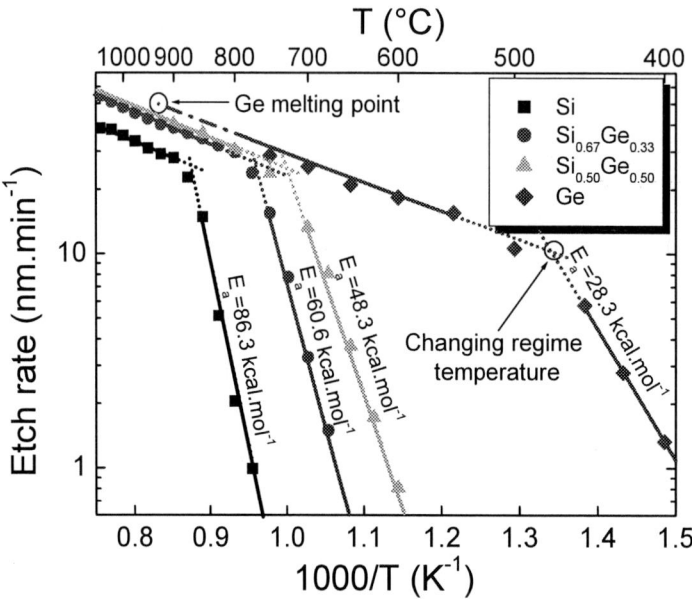

Figure 1: Etching rates of Si, $Si_{0.67}Ge_{0.33}$, $Si_{0.5}Ge_{0.5}$ and germanium as a function the reverse absolute etch temperature, from Y. Bogumilowicz et al (27).

The HCl etch is described as the adsorption of HCl molecules on the Si, SiGe or Ge surface and the consumption of Si or Ge as shown in equations [4] and [5] :

$$Si\ (s) + 2HCl(g) \rightarrow SiCl_2(g) + H_2(g) \qquad [4]$$

$$Ge\ (s) + 2HCl(g) \rightarrow GeCl_2(g) + H_2(g) \qquad [5]$$

The HCl etch depends strongly on the HCl partial pressure in the high temperature regime. However, its dependence on the temperature and on the Ge concentration is weak (low activation energies: ~ 7 kcal.mol^{-1}). By contrast, the HCl etch rate depends strongly in the low temperature regime on the etch temperature (high activation energies: 86 kcal.mol^{-1} for pure Si, 28 kcal.mol^{-1} for pure Ge) and on the Ge content of the layers (27). The mechanism of defects delineation using HCl is not well-known and still under investigation. Some authors have noticed an influence of the strain field distribution on the HCl etch rate (28). The HCl etch pits have been found to be equivalent in terms of densities and distribution with the ones observed after (i) Schimmel on SiGe virtual substrates (27-28), (ii) Secco on strained-silicon layers (27-29) and (iii) Secco and Cr-free etches on pure epitaxial Ge (11). A lateral and anisotropic etching leading to the formation of inverted pyramids with different depths has been evidenced after HCl etches of Ge. By contrast, round and elliptical pits were observed after isotropic chemical etches (Secco and Cr-free). The HCl defect revelation offers several advantages over chemical etches, such as throughput, the processing of full wafers and the lack of mandatory delineation of etch pits afterwards (unlike chemical etches).

Whatever the kind of solution (Cr-based, Cr-free solutions or HCl etch), several studies have been conducted in order to control the etching rate of the reaction, which influences its selectivity. On thin films and new materials, an undoubtful and accurate discrimination between defects is indeed required. To overcome some limitations of old solutions (too fast an etch rate, etching behavior and homogeneity on new materials such SiGe with high Ge content, sSi/SiGe stacks, "on-insulator" and multi-layer substrates etc..), some modifications have first been introduced : a dilution of the solution with water (example : diluted Schimmel (30)), the use of two-step etching on sSi and sSOI (31-32), the etching at low temperature (32-33), the addition of an external agent in the solution, such as NH_4OH in a $HF/HNO_3/CH_3COOH$ - type solution (34).

Differences in surface potential at defect sites is indeed mandatory for selective etching. This etching is not only influenced by the strain field and/or the metallic impurities segregated at the dislocations sites. It is also very dependent on the composition of the solution (or HCl flow and pressure), the temperature adopted for the etching and the structure of silicon or studied material. In a solution, a low temperature can enable the delineation of etch pits by decreasing the etch rate and increasing at the same time the ratio between the substrate etch rate and the defect etch rate. For HCl, the temperature is essentially a function of the substrate which is to be etched.

Some authors have recently compared EBIC with Schimmel on $Si_{1-x}Ge_x$ virtual layers (35). A good defect density agreement has been evidenced over a wide range of Ge contents (defect densities were too small to be quantified thanks to Transmission Electron Microscopy (TEM)). Similarly, Bray et al have investigated the electrical and structural properties of defects in epitaxial Si/SiGe layers thanks to Secco and EBIC (36). They have shown a dependence of defect distribution, stress and defect concentration on the Secco etch rate. Using EBIC measurements at low temperatures, they have studied the electrical activity of dislocations in epitaxial SiGe layers (coming from the charge carrier recombination electrical activity).

The crystallographic orientation, the doping level, the microscopic strain variations, the Ge content of the layer etc are parameters which influence the etching reaction. The nature of the oxidizing agent, its concentration, the ratio between this agent and HF and as well the temperature will determine the oxidation potential of a chemical etch solution. An in-depth knowledge of etching properties is thus needed. We will discuss in the parts that follow the etch behavior of different solutions (Schimmel, Cr-free and Secco etches) and of gaseous HCl as a function of the Ge content of SiGe layers.

Etch rates as a function of the Ge content

Figure 2 shows the $Si_{0.7}Ge_{0.3}$ thickness etched (Ks) as a function of time in a diluted Schimmel and a homemade Cr-free solution composed of $HF/HNO_3/CH_3COOH$. The consumption is linear over time. The slopes of the linear fits of such sets of data represent the etch rates. A higher etch rate is obtained during the Cr-free etch, suggesting a different etching behavior between the Cr-based and the Cr-free etching solutions over time.

The typical etch rates associated to the different etching techniques as a function of the Ge content of the layers have been compiled in Figure 3.

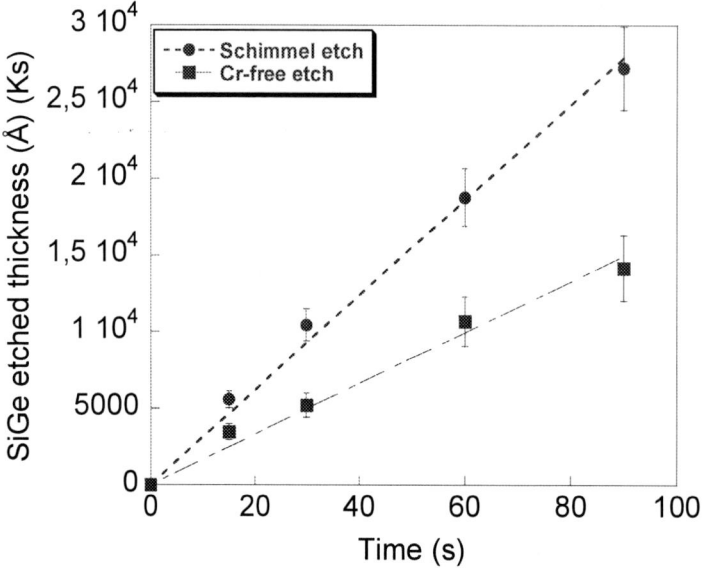

Figure 2: SiGe 30% thickness Ks removed (Ångstroms) as a function of the etching time (seconds).

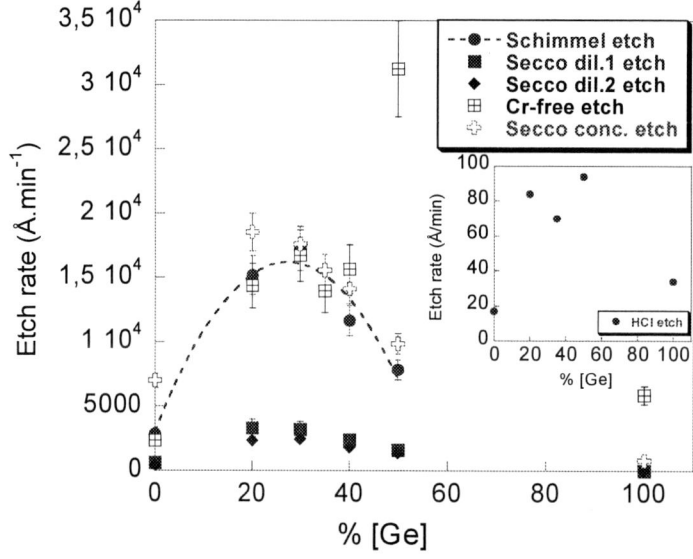

Figure 3 : Etch rates (Å.min⁻¹) of different etching techniques commonly used on Si, SiGe and Ge substrates.

Let us first focus on the wet etching solutions. We do see that the Secco etch rates can be tailored in order to etch different amounts of SiGe for a given dip time. There is indeed a factor 5 difference in etch rates between concentrated and diluted (in water) Secco (1500 vs ~ 300 nm min.⁻¹ for SiGe). A very clear increase of the etch rate occurs for concentrated Secco when switching from Si to $Si_{0.8}Ge_{0.2}$ (from ~ 700 up to 1800 nm min.⁻¹). These results confirm the importance of the oxidizing agent (nature and concentration) on the etching properties. Then, the etch rate monotonously decreases as the Ge content of the layer increases, down to ~ 1000 nm min.$^{-1}$ for $Si_{0.5}Ge_{0.5}$ and less than 100 nm min.$^{-1}$ for pure Ge. The trend is similar, although far less marked, for diluted Secco. As far as Schimmel is concerned, the same bell-like curve is obtained, with an etch rate which increases from ~ 250 nm min.$^{-1}$ for pure Si up to $1500 - 1600$ nm min.$^{-1}$ for $Si_{0.8}Ge_{0.2}$ and $Si_{0.7}Ge_{0.3}$. It then decreases monotonously down to ~ 800 nm min.$^{-1}$ for $Si_{0.5}Ge_{0.5}$ and virtually 0 for pure Ge. The behavior with the Cr-free solution we have developed for defect revelation is somewhat different. We have indeed an increase from ~ 200 nm min.$^{-1}$ up to $1400 - 1600$ nm min.$^{-1}$ when moving from Si to $Si_{1-x}Ge_{x}$, with x being equal either to 20%, 30% or 40%. The etch rate unexpectedly jumps up to 3200 nm min.$^{-1}$ for $Si_{0.5}Ge_{0.5}$ (with a large error bar, however), before falling down to a ~ 600 nm min.$^{-1}$ value for pure Ge. Such an etch rate is much higher than the one obtained with concentrated Secco or Schimmel.

In Cr-based solutions, an adsorption of Cr on Si and Ge atoms via Si-OH bonds occurs. Electron transfer then happens through bridged atoms, independently of the form of hexavalent chromium (CrO_3 or $K_2Cr_2O_7$, provided that the degree of oxidation of Cr^{6+} is respected). The more the Ge concentration increases, the less strong bonds are, reducing the adsorption of Cr on the surface and consequently the electron transfer, explaining the reduction in etch rate. Different etching mechanisms are at play with Cr-free etches. As described in the first part of the article, electrochemical reactions between the oxidizing agent (HNO_3) and the Si surface are necessary. As Ge atoms are incorporated in the SiGe layers, the perfect Si lattice becomes more and more vulnerable to oxidizing agents such as H_2O_2 and to a lesser extent HNO_3. Depending on the ratio HF/HNO_3 ratios, such oxidizing agents might act more and more as dissolving etchants, as more Ge atoms are exposed to the etchant (and thus oxidized) (37). Such conclusions have already been drawn by several authors when dealing with the etching of SiGe in $HF/H_2O_2/CH_3COOH$ solutions (38-39). As Si is hardly oxidized by HNO_3 (40-41), those authors concluded that the increase in etch rate with the Ge content could be understood by an increased number of Ge atoms. It should finally be highlighted that the bond strength of Si-Si is higher (77.5 kcal.mol.$^{-1}$) than the bond strengths of Si-Ge (~ 71 kcal.mol.$^{-1}$) and Ge-Ge (63 kcal.mol.$^{-1}$).

Defects revelation in SiGe and sSi/SiGe stacks

We have estimated using an optical microscope in the Dark Field mode, the $Si_{0.7}Ge_{0.3}$ etch pit size (represented by the length of the axis of the isotropic elliptical pits) evolution as a function of the etching time (see Figure 4). An increase of the defect size Kd with the amount Ks of SiGe removed is suggested from observations. The selectivity of Schimmel and Cr-free solutions is deduced from the Kd/Ks ratio. The Cr-free etch presents a higher selectivity than the Schimmel etch. The selectivity of gaseous HCl etching is comparatively far higher (>20 instead of ~ 1.4 for Cr-free and <1 for Schimmel), leading to cristallographically oriented pits, as shown in Figure 5. Some optical microscope images after a Schimmel etch and a HCl etch of $Si_{0.7}Ge_{0.3}$ are shown in Figure 6 and 7 (the image after a Cr-free etch is very similar to the one after a Schimmel etch).

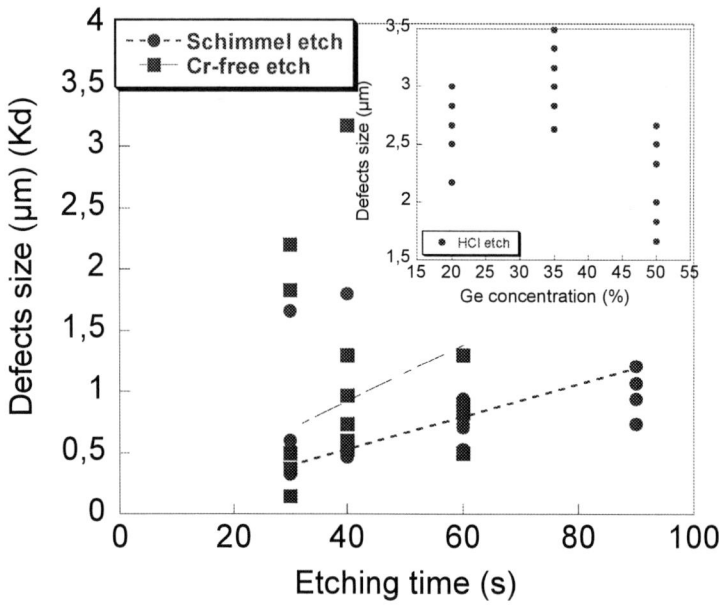

Figure 4 : Evolution of defects size Kd as a function of time (s) after a Schimmel etch, Cr-free etch and a gaseous HCl etch on a $Si_{0.7}Ge_{0.3}$ virtual substrate.

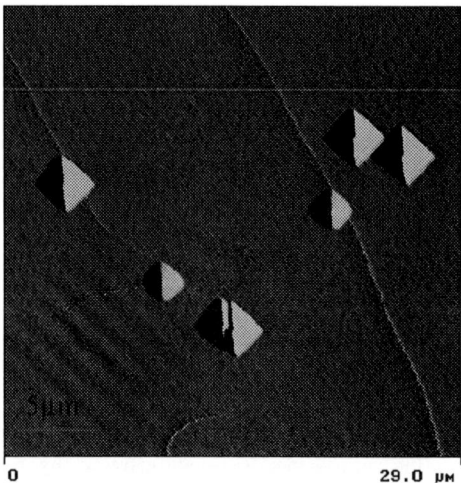

Figure 5 : Amplitude mode Atomic Force Microscopy image of the surface of a $Si_{0.65}Ge_{0.35}$ virtual substrate after the gaseous HCl etch of 56 nm of SiGe. Image sides are more or less along the <100> directions.

Figure 6 : (130x100)μm^2 optical microscopy image after a Schimmel etch of a $Si_{0.7}Ge_{0.3}$ virtual substrate.

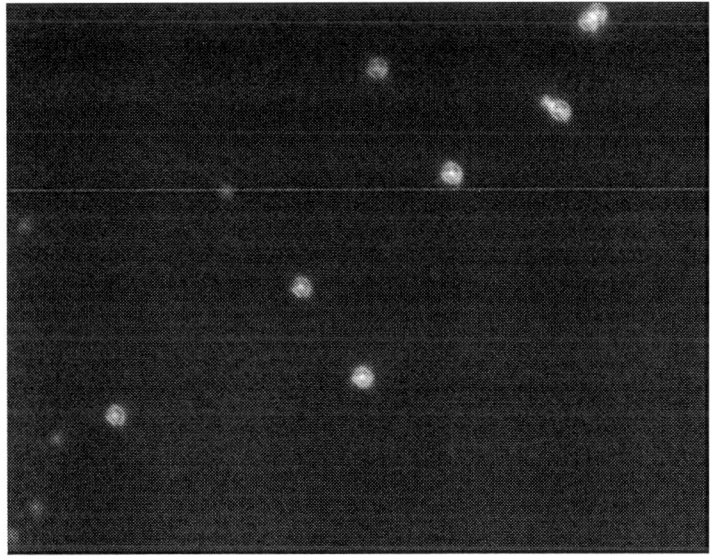

Figure 7 : (130x100)μm^2 optical microscope image after a HCl etch of a $Si_{0.7}Ge_{0.3}$ virtual substrate.

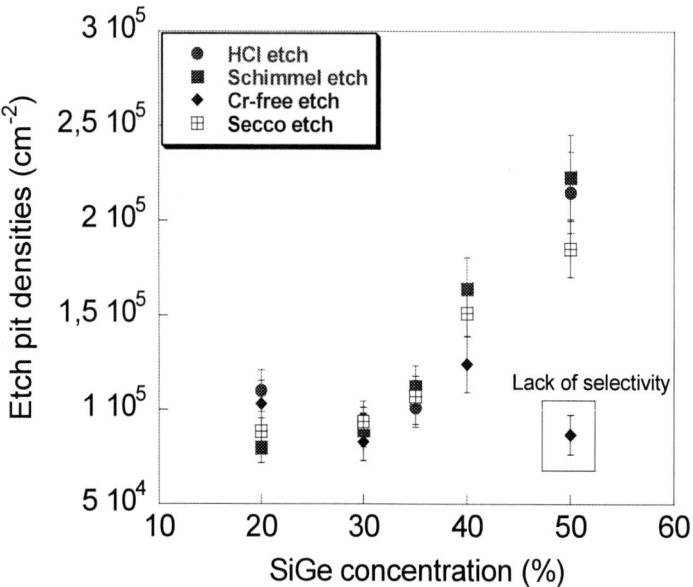

Figure 8 : Etch pit density (cm^{-2}) as a function of the Ge content of the virtual SiGe substrates for different etches.

Finally, we have compared the Etch Pit Densities (EPD) as a function of the Ge content of the SiGe substrates, after different etches (see Figure 8). A really good agreement is obtained between the miscellaneous defect revelation techniques, meaning that the actual Threading Dislocations Densities (TDD) must be more or less equal to the EPD. The TDD increases from values in the high $10^4 - 10^5$ cm^{-2} range for Ge contents around 20% to 25% to values closer to 2×10^5 cm^{-2} for a Ge content equal to 50%. This is most likely due to the surface roughening occurring for higher Ge contents (9) that somewhat inhibit the glide of pre-existing threading dislocations, leading to the formation of new dislocations (hence the higher TDD). An EPD underestimation is noticed on $Si_{0.5}Ge_{0.5}$ after a Cr-free etch. Such a trend is likely due to a selectivity loss of the Cr-free solution on high Ge content SiGe virtual substrates. An etching with a modified solution or the use of low temperature could potentially be advantageous for a better delineation of etch pits in high Ge content SiGe substrates.

The selectivity of Secco has already been reported to be higher than Cr-O$_3$ based chemistries on Si substrates (28). We have largely extended the use of Secco etch (commonly used on SOI) to (i) sSOI and XsSOI monitoring (4) (29) and (ii) sSi/SiGe stacks monitoring (such stacks indeed constitute the starting material for sSOI elaboration). The high selectivity of Secco (>2) enables the delineation of different types of defects: 60° threading dislocations, pile-ups, 90° dislocations half-loops and micro-twins (42). It is thus a routine technique, which can notably be used for the estimation of TDDs in strained silicon layers grown on top of SiGe virtual substrates (see Figure 6). What is easily seen when comparing Figure 9 data with the ones of Figure 5 is that the TDD (in the $9 \times 10^4 - 3 \times 10^5$ cm^{-2} range) is apparently dictated by the TDD of the SiGe virtual substrate underneath, even if the interface between SiGe and sSi has been demonstrated to be important for a good quality sSi growth (43-46). Almost no variation of the in-plane tensile strain with the sSi layer thickness (in-between 5 and 37 nm) was evidenced in UV-Raman for sSi layers grown on top of miscellaneous Ge content SiGe virtual substrates (Figure 10). The tensile-strain decay slopes are indeed equal to $\sim$ 0, 0.1 and 1.1 MPa nm^{-1} only for sSi layers grown with SiH$_2$Cl$_2$ at 700°C on Si$_{0.79}$Ge$_{0.21}$, Si$_{0.71}$Ge$_{0.29}$ and Si$_{0.60}$Ge$_{0.40}$ VS, respectively. Only for sSi layers grown on Si$_{0.52}$Ge$_{0.48}$ VS does the tensile strain decay slope increases significantly to a 6.0 MPa nm^{-1} value.

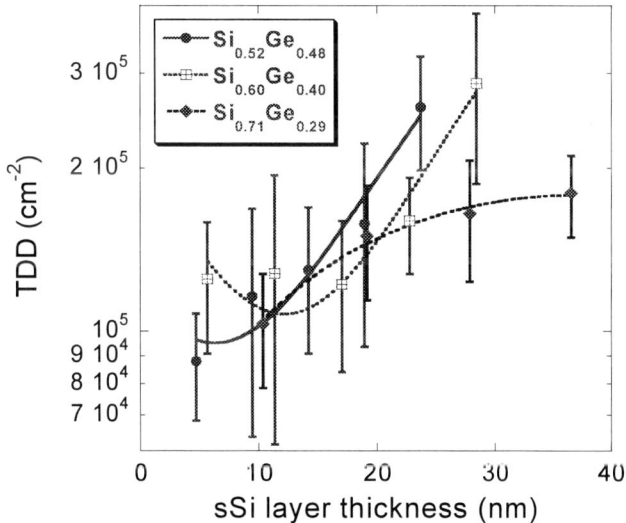

Figure 9 : Threading dislocations densities as a function of the sSi thickness grown on top of SiGe virtual substrates with different Ge contents (Secco).

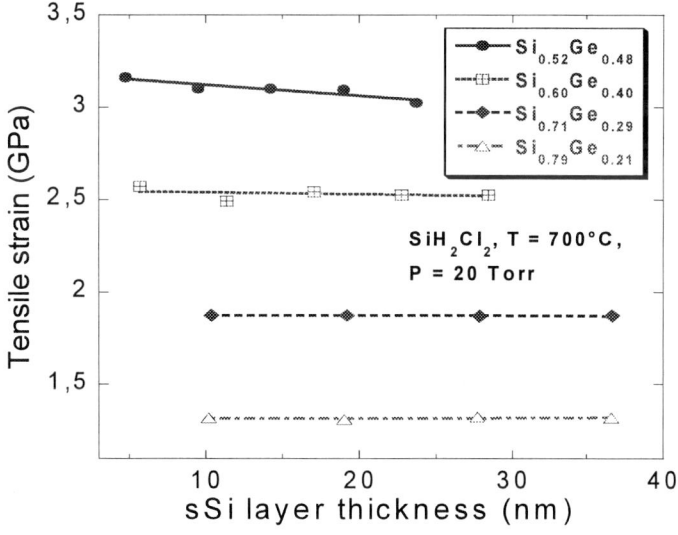

Figure 10 : Tensile-strain in the sSi layers grown on top of Si$_{0.52}$Ge$_{0.48}$, Si$_{0.60}$Ge$_{0.40}$, Si$_{0.71}$Ge$_{0.29}$ and Si$_{0.79}$Ge$_{0.21}$ virtual substrates, this as a function of their thickness.

We can thus quite safely state that the sSi layers are fully pseudomorphic in the full {sSi layer thickness, Ge concentration} domain probed in this study, despite the fact that the theoretical values are slightly higher than the ones found in UV-Raman. Crystalline quality analysis using Secco becomes therefore the only reliable method to discriminate between different {SiGe/sSi} substrates. It can also be used as a routine technique on sSi/SiGe stacks for the qualification of tools such as epitaxy reactors, in complement of others techniques such as Secondary Ion Mass Spectrometry (SIMS).

Finally, we have compiled in Figure 11 the selectivity of the different etching techniques commonly used (defined as the Kd/Ks ratio). As mentioned before, the HCl selectivity is a factor 10 higher than the one of wet etching solutions (40 versus 2-3 for Si, 40 versus 1 for SiGe and 9 versus less than 1 for pure Ge).

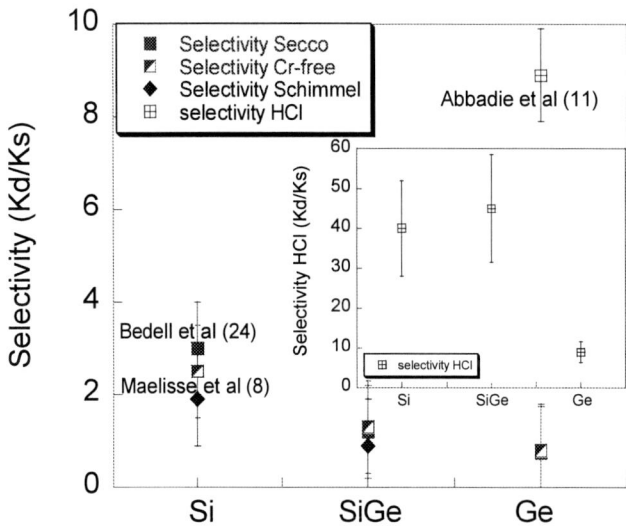

Figure 11 : Selectivity of different etching techniques as a function of the type of substrate from Si to Ge.

Recent developments in Cr-free etches

A new Cr-free etch solution (named FS Cr-free SOI) has recently been developed for the delineation of crystalline defects in SOI, in order to anticipate future restrictions in Cr^{6+} use (8). It consists of a mixture of $HF/HNO_3/CH_3COOH$, in which the ratio of the components has been tested and adjusted accurately. The particularity of a such solution compared to the common Cr-free solutions described in the first paragraph of this article, is the presence of bromine (which is added to the solution). Bromine is the key factor in this solution and is absolutely necessary for the etching on SOI as it initiates the oxidation reaction which is then carried out by nitric acid. A performance similar to a diluted Secco etching solution commonly used as a standard has been evidenced. An homogeneous surface is produced afterwards with well-developed etch pits at defect sites (Figure 12). After a subsequent HF step (also used after diluted Secco etches), these defects can be clearly seen with an optical microscope.

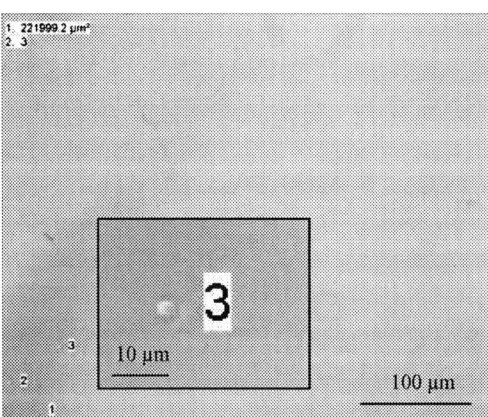

Figure 12 : Optical microscope image (2.22x10⁵ μm²) of a the surface of a SOI substrate etched down to 300Å after a "two-step" etching in the FS Cr-free SOI solution, from J. Maelisse et al (8).

An excellent correlation with the Secco solution has been found. As far as the oxidation is concerned, several authors suggested a mechanism based on an interaction between silicon, bromine and hydrofluoric acid (47-48). After removal of the native oxide layer by fluoride, the exposed silicon atoms react with bromine to form two Si-Br bonds (step 1). The polarized Si-Br bonds increase the chemical reactivity of the other Si back bonds which are, therefore, more easily attacked by HF (step 2). Finally, the activated Si-H bonds, which are also unstable in aqueous and fluoride media, will undergo nucleophilic attack (involving hydrogen or the decomposition products of HNO_3 if present in the solution) (step 3). The mechanism is shown Figure 13.

Figure 13 : Oxidation mechanism of silicon by bromine in the absence of nitric acid.

The bromine is a strong oxidizing agent [equation 6].

$$Br_2 \text{ (aq)} + 2e\text{-} \leftrightarrow Br^- \qquad E^o (Br_2/Br^-) = 1.08 \qquad [6]$$

For safety and handling reasons, it has been replaced in the solution by salts of bromate and bromide. In acid solutions, bromate and bromide generate bromine according to equation [7]:

$$BrO_3^- + 5\,Br^- + 6\,H^+ \rightarrow 3\,Br_2 + 3\,H_2O. \qquad [7]$$

The bromine content influences the etch rate of the solution. Some pseudo-MOS measurements on SOI samples after etching in the FS Cr-free solution seem to confirm the adsorption of bromine on the Si surface thanks to the formation of Si-X termination bonds. We have indeed observe a negative shift of the flat band potential Vfb which increases with the bromine content. Such observations have already been reported by several authors on Si(111) surfaces (48-49). We assume that this modification of the Si surface bonds is linked to a modification of the surface potential. Such results will be detailed in another publication.

Another class of preferential etch solutions very recently proposed are the *organic peracid etches* (7). These are free of toxic Cr^{6+} species and contain no nitric acid. They consist of hydrofluoric acid, an organic acid (e.g. acetic acid or a propionic acid) and hydrogen peroxide which reacts with the organic acid forming the organic peracid after a few hours. The organic peracid acts as the oxidizing agent. Such solutions are characterized by very slow etch rates (0.4 –1.5 nm/min at 25°C) and present a very high selectivity for defects delineation. The first observations have been performed by Possner et al. on Cz bulk silicon (7). This solution would enable the differenciation between hillocks and vacancies (the last ones appearing as octahedral-shaped etch pits).

Conclusions

Three kind of etching techniques are used nowadays: Cr-based solutions, Cr-free etch solutions and gaseous HCl etchings. The Cr-based solutions, such as Secco, Schimmel or Wright, are widely used for defects delineation in Si, SiGe and Ge. Thanks to the high oxidizing power of chromium, these solutions enable a high delineation of defects. They are used routinely as monitoring techniques on SOI, sSOI and GeOI. One of the limitations of such techniques, besides the high toxicity of chromium, is often the high etch rates which are inherent to such solutions. Therefore, some modifications of existing solutions are needed to adapt to new materials and thin films. Moreover, the mechanism of etching is not really known.

Cr-free etch solutions are commonly based on the use of acid hydrofluoric, nitric acid and acetic acid. Nitric acid constitutes the oxidizing agent. HF is necessary to dissolve the oxide formed by HNO_3 and its decomposition products. Acetic acid (CH_3COOH) is the dissolving agent. The main difficulty with such solutions lies in the fact that the ratio between HF/HNO_3 needs to be adapted to each kind of material. The nature of the substrate (crystalline orientation, doping, Ge content..) are parameters to be considered. An example of Cr-free development has been described in this article for SiGe virtual substrates etching. An etching mechanism different from the one of Secco or Schimmel has been evidenced. Anyway, a satisfying correlation is obtained between all techniques. The same kind of correlation has recently been published on pure Ge wafers. However, on "on-insulators" substrates, some modifications of these solutions are needed to obtain an homogeneous and selective etching. Some additional oxidizing agents such Br⁻, I⁻…are indeed necessary to initiate the oxidation process which then is carried out by nitric acid. Such mechanisms are currently studied more in depth.

In order to increase the selectivity of such wet-etches solutions (Cr-based or not), ones has either to use a HF delineation step afterwards or lower temperatures. The gaseous HCl etch technique, studied these last two years, has the advantage of delineating defect etch pits much better. We have used it on Si, SiGe, Ge, SOI, sSOI and GeOI. No further etch pits delineation is needed after etching and a very good selectivity is obtained (>20 versus close to 1-2 for wet-etch solutions). Its main limitation (or its advantage) is that a full wafer processing in a clean-room is required.

Acknowledgements

The author would like to acknowledge Pr B.O. Kolbesen, J. Maelisse and D. Possner from Frankfürt University, Germany, for their active participation. All persons from CEA-LETI and SOITEC having participated to this work are gratefully acknowledged.

References

1. B. Ghselen, J.M. Hartmann, T. Ernst, C. Aulnette, B. Osternaud, Y. Bogumilowicz, A. Abbadie, P. Besson, O. Rayssac, A. Tiberj, N. Daval, I. Cayrefourcq, F. Fournel, H. Moriceau, C. Di Nardo, F. Andrieu, V. Paillard, M. Cabié, L. Vincent, E. Snoeck, F. Cristiano, A. Rochet, A. Poncet, A. Claverie, P. Boucaud, M.N. Semeria, D. Bensahel, N. Kernevez and C. Mazure, S*olid State Electron.* **48**, 1285 (2004) 1285.
2. F. Letertre, C. Deguet, C. Richtarch, B. Faure, J.M. Hartmann, F. Chieu, A. Beaumont, J. Dechamp, C. Morales, F. Allibert, P. Perreau, S. Pocas, S. Personnic, C. Lagahe-Blanchard, B. Ghyselen, Y.M. Le Vaillant, E. Jalaguier, N. Kernevez and C. Mazure, *Mat. Res. Soc. Symp. Proc.* B4.4.1, 809 (2004).
3. I. Cayrefourcq, M. Kennard, F. Metral, C. Mazuré, A. Thean, M. Sadaka, T. White and B.Y. Nguyen, in *Silicon-on-Insulator Technology and Devices XII,* ECS Proc. Vol.2005-03 (2005), edited by G. Celler, S. Cristoloveanu, F. Gamiz, J.G. Fossum and K. Izumi, Editors, (The Electrochemical Society, Pennington, NJ, USA, 2005), pp.191-206.
4. T. Akatsu, J.M. Hartmann, A. Abbadie, C. Aulnette, Y.M. Le Vaillant, D. Rouchon, Y. Bogumilowicz, L. Portigliatti, C. Colnat, N. Boudou, F. Lallement, F. Triolet, C. Figuet, M. Martinez, P. Nguyen, C. Delattre, K. Tsiganenko, C. Berne, F. Allibert, C. Deguet, M. Kennard, E. Guiot, F. Metral and I. Cayrefourcq, *ECS Transactions* **3** (6), 107 (2006).
5. A. Virtuani, S. Marchionna, M. Acciarri, G. Isella and H. Von Kaenel, *Mat. Sci. Semicond. Proc.* **9**, 798 (2006).
6. J.M. Hartmann, A. Abbadie, D. Rouchon, Y. Guinche, P. Holliger, G. Rolland, M. Buisson, F. Brunier, C. Defranoux, F. Pierrel and T. Billon, *ECS Transactions* **3** (7), 309 (2006).
7. D.Possner, B.O.Kolbesen, V.Klüppel and H.Cerva, ALTECH Symp., Munich Sep 2007, to be published in ECS Transaction, Oct 2007.
8. J. Mähliß, A. Abbadie and B. O. Kolbesen, *ECS Trans.* **6** (4), 271 (2007).
9. Y. Bogumilowicz, J.M. Hartmann, F. Laugier, G. Rolland, T. Billon, N. Cherkashin and A. Claverie, *J. Cryst. Growth.* **283**, 346 (2005).
10. J.M. Hartmann, J.F. Damlencourt, Y. Bogumilowicz, P. Holliger, G. Rolland and T. Billon, *J. Cryst. Growth* **274**, 90 (2005).
11. A. Abbadie et al., *ECS Trans.* **6** (4), 263 (2007).
12. E. Sirtl and Adler, *Z. Metallk.* **52**, 529 (1961).
13. M.W. Jenkins, *J. Electrochem. Soc.* **124**, 757 (1977).
14. F. Secco d'Aragona, *J. Electrochem. Soc.* **119**, 948 (1972).
15. D.G. Schimmel, *J. Electrochem. Soc.* **126**, 479 (1979).
16. R.W. Maatman and A. Kramer, *J. Phys. Chem.* **72**, 104 (1968).
17. W.C. Dash, *J. Appl. Phys.* **27**, 1193 (1956).
18. H. Robbins and B. Schwartz, *J. Electrochem. Soc.* **107**, 2 (1960).

19. Y. Kashigawa, R. Shimokawa and M. Yamanaka, *J. Electrochem. Soc.* **143**, 2 4079 (1996).

20. M. Steinert, J. Ackert, A. Henaige and K. Wetzig, *J. Electrochem. Soc.***152**, 12 C843 (2005).

21. J. Ackert, M. Steinert, A. Henaige and K. Wetzig, *Meet. Abstract Electrochem. Soc.* **501**, 742 (2006).

22. S. Verhaverbeke, I. Steerlinck, C. Vinckier, G. Stevens, R. Cartuyvels and M.M. Heyns, *J. Electrochem. Soc.* **141**, 10 2852 (1994).

23. M. Kittler, C. Ulhaq-Bouillet, J. Hersener and F. Schaeffler, *Solid. St. Phenom.* **32**, 559 (1993).

24. V. Higgs and M. Kittler, *Inst. Phys. Conf. Ser.*, **146**, 723 (1995).

25. T.C. Chandler, *J. Electrochem. Soc.* **137**, 944 (1990).

26. "MEMC Electronic Materials Standard Evaluation Test Methods", STD 500-01, p.4, MEMC Electronic Materials Co., Inc., St. Louis, Missouri (1989).

27. Y. Bogumilowicz, J.M. Hartmann, R. Truche, Y. Campidelli, G. Rolland and T. Billon, *Semicond. Sci. Techn.* **20**, 127 (2005).

28. S.W. Bedell, H. Hovel, A. Domenicucci, K. Fogel, A. Reznicek and D.K. Sadana, *Silicon-on-Insulator Technology and Devices XII,* ECS Proc. Vol.**2005-03** (2005), edited by G. Celler, S. Cristoloveanu, F. Gamiz, J.G. Fossum and K. Izumi, Editors, (The Electrochemical Society, Pennington, NJ, USA, 2005), pp.345-356.

29. Abbadie, S.W. Bedell, J.M. Hartmann, D.K. Sadana, F. Brunier, C. Figuet and I. Cayrefourcq, *J. Electrochem. Soc.* **154**, H713 (2007).

30. J. Werner, K. Lyutovich and C.P. Parry, *Eur. Phys. J. Appl. Phys.* **27**, 367 (2004).

31. W.Bedell, H.Chen, D.K.Sadana, K.Fogel and A.Domenicucci, *Electrochem. & Solid State Lett.* **7**, G105-G107 (2004).

32. J.Lu, R.Zhang, G.Rozgonyi, E.Yakimov and N.Yarykin, *ECS Transactions* **2** (2), 569 (2006)

33. K.R. Bray, W. Zhao, L. Kordas, R. Wise, McD. Robinson and G. Rozgonyi, *J. Electrochem. Soc.* **152**, C310 (2005).

34. Z. Liu, T. Sun, J. An, J. Wang, X. Xu and R. Cui, *J. Electrochem. Soc.* **154**, D21 (2007).

35. A. Virtuani, S. Marchionna, M. Acciarri, G. Isella and H. Von Kaenel, *Mat. Sci. Semicond. Proc.* **9**, 798 (2006).

36. K.R. Bray, W. Zhao, L. Kordas, McD. Robinson and G. Rozgonyi, *J. Electrochem. Soc.* **152**, C310 (2005).

37. A. Abbadie, J.M. Hartmann, C. Di Nardo, T. Billon, Y. Campidelli and P. Besson, *Microelec. Eng.* **83**, 1986 (2006).

38. G.K. Chang, T.K. Carns, S.S. Rhee and K.L. Wang, *J. Electrochem. Soc.* **138**, 202 (1991).

39. T.K. Carns, M.O. Tanner and K.L. Wang, *J. Electrochem. Soc.* **142**, 1260 (1995).

40. B. Schwartz and H. Robbins, *J. Electrochem. Soc.* **111**, 196 (1964).

41. W. Kern and D.A. Puotinen, *RCA Rev.*, 187 (1970).

42. O. Kononchuck, F. Brunier and M. Kennard, *ECS Trans.* **6** (4), 225 (2007).

43. C. Leitz, V. Yang, M. Carroll, T. Langdo, R. Westhoff, C. Vineis, M. Bulsara, *Mat. Sci. Semicond. Proc.* **8**, 187 (2005).

44. M.S Carroll, J.C. Sturm, M. Yang, *J. Electrochem. Soc.* **147**, 4652 (2000).

45. A. Abbadie, J.M. Hartmann, P. Holliger, M.N. Semeria, P. Besson, P. Gentile, *Appl. Surf. Sci.* **225**, 256 (2004).

46. C.W. Leitz, C.J. Vineis, J. Carlin, J. Fiorenza, G. Braithwaite, R. Westhoff, V. Yang, M. Carroll, T.A. Langdo, K. Matthews, P. Kohli, M. Rodder, R. Wise, A. Lochtefeld, *Thin Solid Films* **513**, 300 (2006).
47. S. Meltzer and D. Mandler, *J. Chem. Soc. Faraday Trans.* **91** (6), 1019 (1995).
48. X. Zhou, M. Ihida, A. Imanishi and Y. Nakato, *Electrochemica Acta*, **45**, 4655 (2000).
49. B.J. Eves and G.P. Lopinski, *Surf. Sci. Lett.* **579**, L89 (2005).

Organic Peracid Etches: a new class of chromium free etch solutions for the delineation of defects in different semiconducting materials

D. Possner[a], B.O. Kolbesen[a], H. Cerva[b], V. Klüppel[b]

[a] Department of Inorganic and Analytical Chemistry, J.W.G. University,
D-60438 Frankfurt am Main, Germany
possner@chemie.uni-frankfurt.de, kolbesen@chemie.uni-frankfurt.de
[b] Corporate Technology Analytics, Siemens AG, D-81730 Munich, Germany

Defect etching is a well-established method used to reveal different kinds of crystalline defects in semiconducting materials. Most of the etch solutions used today have two disadvantages. They contain hexavalent chromium which is highly toxic and they are not suitable for application on thin films. There is a demand for environmentally friendly etch solutions which can also be used for new materials like SOI. Due to their properties Organic Peracid Etches (OPE), mixtures which contain a short-chain alkanoic acid like acetic or propionic acid, hydrogen peroxide and hydrofluoric acid are suitable for defect delineation in thin and very thin (<50 nm) films. Such solutions were also tested on CZ and FZ silicon substrates. In this case characteristic square-shaped etch figures caused by D-defects (COPs) were found after etching.

New etch solutions for new materials

Novel substrate materials like Silicon On Insulator (SOI), Strained Silicon On Insulator (sSOI) or Germanium On Insulator (GeOI) have already been introduced (SOI) or are candidates (sSOI, GeOI) to overcome the limitations of conventional silicon substrates or epitaxial wafers regarding further progress in the performance of microelectronic devices (1). SOI consists of a thin layer of silicon on top of an insulator like silicon dioxide (SiO_2). Starting from silicon bulk SOI can be produced by the Smart Cut ® or Simox process (2). For quality control of SOI wafers a diluted version of the Secco etch (3) is normally used. The Secco diluted solution contains $K_2(Cr_2O_7)$ and HF. Chromate (Cr(VI)) ions are highly toxic. Due to their non-uniform etch removal and delamination of the SOI layer, existing chromium free recipes like the Dash solution (4) are not suitable for application on thin and very thin films. In this work the Dash solution was modified in several ways to adapt it to thin films. The most effective results were obtained when the nitric acid was replaced by an equal volume of hydrogen peroxide.

	HNO_3 (65%)	HF (49%)	Hac (100%)
Dash recipe:	43 ml	14.5 ml	143 ml

	H_2O_2 (30%)	HF (49%)	Hac (100%)
Modified Dash recipe: ("Organic peracid etch (OPE) A")	43 ml	14.5 ml	143 ml

Figure 1: Original and modified Dash recipe (OPE A)

This new etch solution based on the Dash recipe is called Organic Peracid Etch A (OPE A) because mixtures of acetic acid and hydrogen peroxide always contain peracetic acid (5), equation 1.

$$H_3C\text{-}COOH + H_2O_2 \longrightarrow H_3C\text{-}COOOH + H_2O \qquad [1]$$

The peracetic acid is assumed to be the active species which oxidizes the silicon (equation 2). After the oxidation the silicon dioxide is dissolved by hydrofluoric acid (equation 3).

$$Si + 2\ H_3C\text{-}COOOH \longrightarrow SiO_2 + 2\ H_3C\text{-}COOH \qquad [2]$$

$$SiO_2 + 6\ HF \longrightarrow H_2SiF_6 + 2\ H_2O \qquad [3]$$

OPE A has a very low removal rate (0.6 nm/min, 25°C) which compared to Secco solution is about a factor 100 lower. It is capable of making defects visible. It was tested on different SOI materials and the defects revealed were of round shape (figure 2). Detailed investigations by Atomic Force Microscopy (AFM) show that the etch figures are hillocks (figure 3 and 4).

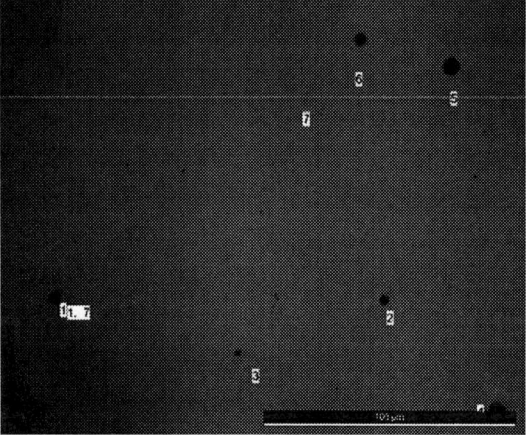

Figure 2: Etch hillocks in SOI material found after etching with OPE A.
Optical micrograph

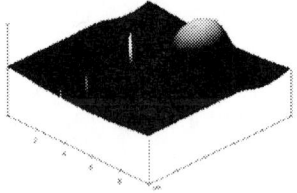

Figure 3: AFM image of an etch hillock (scan area 10 x 10 μm).

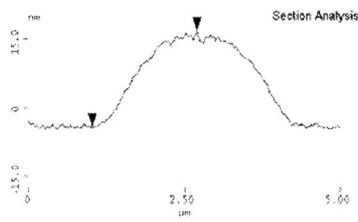

Figure 4: AFM linescan (height of hillock 15 nm, extension 5 μm)

The OPE A is very sensitive for defects. Compared to the Secco diluted reference the defect densities which were found after etching were from 10 to twenty times higher.

Influence of hydrogen peroxide content

To investigate the influence of the hydrogen peroxide content six different etch solutions based on the OPE A were prepared (table 1). The removal rates of these new recipes were determined at room temperature (25°C) on different SOI materials. The hydrogen peroxide and the peracetic acid (PAA) contents were determined by iodometry (6).

Table I. Composition and properties of the different Organic Peracid Etches (OPE)

Composition	OPE A1	OPE A	OPE B	OPE A2	OPE A3	OPE C
HF (50%)	14.5 ml	14.5 ml	43 ml	14.5 ml	14.5 ml	50 ml
H_2O_2 (30%)	30 ml	43 ml	43 ml	72.5 ml	87 ml	-
H_2O_2 (50%)	-	-	-	-	-	50 ml
Hac (100%)	156 ml	114 ml	114 ml	116 ml	102 ml	100 ml
c H_2O_2 + c PAA (mol/l)	1.43	2.14	2.17	3.7	4.37	4.4
c PAA (mol/l)	1.17	1.56	1.6	2.04	1.88	2.85
Removal rate (nm/min)	0.27	0.59	0.54	1.08	1.14	1.34

The removal rate was directly proportional to the total concentration of hydrogen peroxide and peracetic acid (Figure 5). All of these solutions are in principle able to reveal defects in SOI material. Etch solutions with a high hydrogen peroxide content produce a very inhomogeneous removal. Also the size of the etch figures decreases.

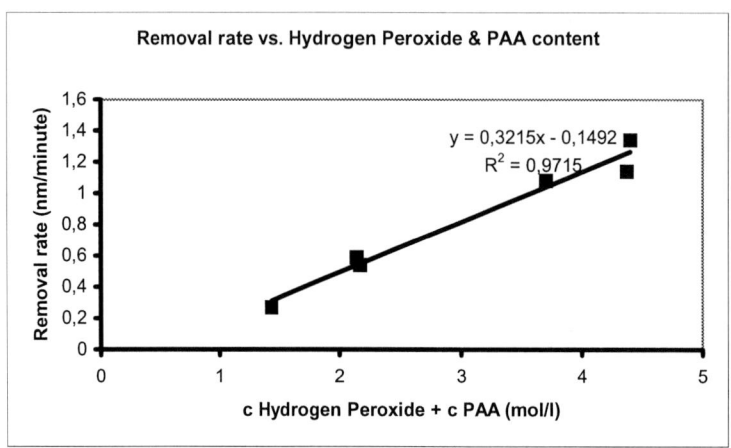

Figure 5: Removal rate dependence on the total concentration of hydrogen peroxide and peracetic acid.

Influence of temperature

The influence of temperature was also investigated. The removal rates were determined at four different temperatures viz. 15°C, 25°C, 35°C and 45°C and found to increase with increasing temperature within this range.
The increase is exponential obeying the Arrhenius law as expected.
The figure below shows the dependence of the removal rate on temperature for the OPE C solution.

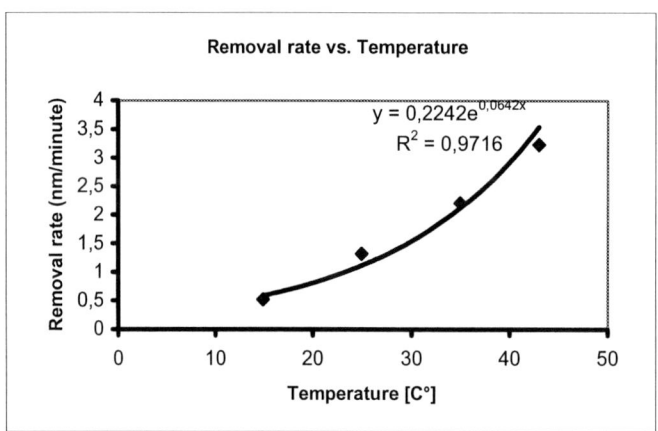

Figure 6: Removal rate dependent on temperature

Etch solutions containing other alkanoic acids

The acetic acid can be replaced by other alkanoic acids like formic-, propionic-, or butyric acid. These organic acids also form peracids with hydrogen peroxide.
Based on the OPE C solution the acetic acid was replaced by formic-, propionic-, and butyric acid in the ratio 1:1. All of these new solutions have similar properties and all of them are able to reveal defects in SOI material as evident from Table II.

TABLE II. Composition and properties of the OPE C – F solutions

Etch solution	Composition			Removal rate (nm/min) at 25°C
OPE C	50 ml H_2O_2 (50%)	50 ml HF (50%)	100 ml Acetic acid (100%)	3.67
OPE D	50 ml H_2O_2 (50%)	50 ml HF (50%)	100 ml Propionic acid (99%)	1.32
OPE E	50 ml H_2O_2 (50%)	50 ml HF (50%)	100 ml Butyric acid (100%)	1.75
OPE F	50 ml H_2O_2 (50%)	50 ml HF (50%)	100 ml Formic acid (100%)	1.74

Etch solutions which contain performic acid are not suitable for practical use because performic acid is very instable, decomposing into carbon dioxide and water (5).
Due to the unpleasant odour of butyric acid the OPE E is also not suitable for practical use.

Influence of the hydrofluoric acid content

Hydrofluoric acid is the third component of the Organic Peracid Etches. Three different etch solutions with increasing HF content based on the OPE A solution were prepared.
An increase in HF content has no influence on removal rate but on the diameter of the etch figures which increase (Figure 7).

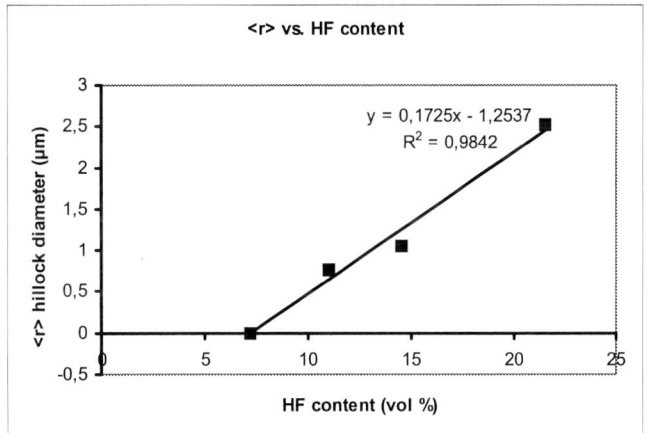

Figure 7: Dependence of the size of the etch figures on HF content.

Nature of the etch hillocks

Detailed SEM investigations show a little hole in the center of the etch hillocks that penetrates the entire silicon layer. It appears that this little hole corresponds to the original defect (Fig.8).

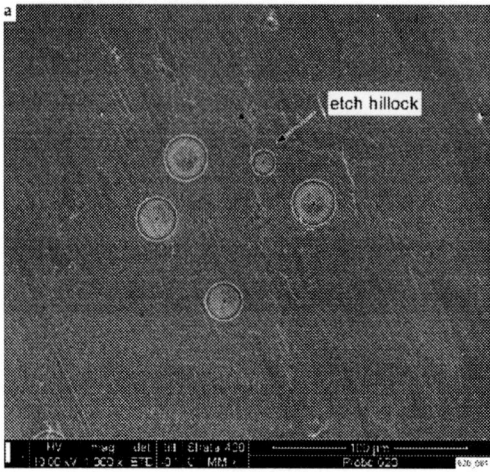

Figure 8: SEM image of etch hillocks in SOI material found after treatment with the OPE D solution.

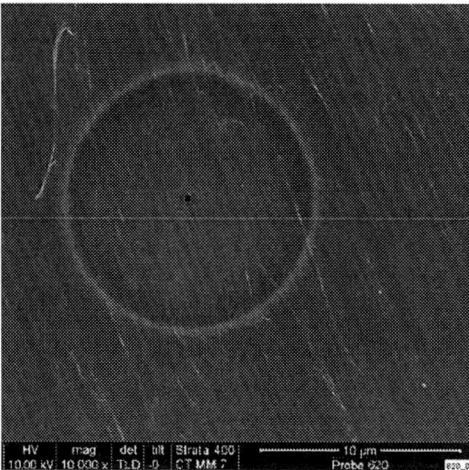

Figure 9: SEM image of an etch hillock. The little hole in the center is easily recognizable.

Further investigations of these etch hillocks by SEM and TEM on cross sections revealed an underlying cavity (figures 10 and 11). Hydrofluoric acid penetrates the hole in the center of the hillock and dissolves the underlying Buried Oxide (BOX) isotropically. This should explain the increase in the size of the hillock with increasing HF content of the etch solution. Later the thin silicon film above starts to bow up and a hillock is formed.

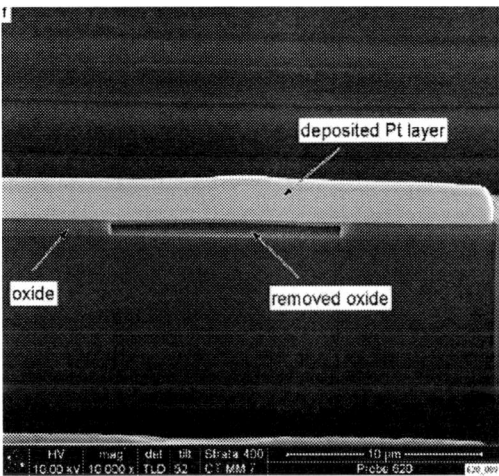

Figure 10: SEM image of an etch hillock cut by a Focused Ion Beam (FIB).

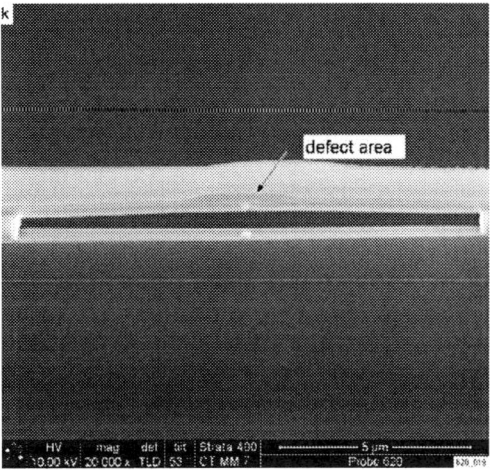

Figure 11: XTEM image of an etch hillock showing the location of the original defect, the underlying cavity and bow up of the Si film.

Organic Peracid Etches tested on bulk material

Microdefects in silicon related to intrinsic point defects were first observed in the early 1960s (7). There are two different kinds of atomic point defects, vacancies and Si self-interstitials, which can occur in silicon substrates and may give rise to the formation of microdefects (8). The interstitial type defects can be classified as A and B defects. A defects are extrinsic dislocation loops. The B defects seem to be small clusters of interstitials. The incorporation of point defects depends on parameters like the pulling speed (V) and axial temperature gradient (G) near the growth interface.
The critical V/G ratio is about 0.2 mm^2/min K. Crystals grown with a higher ratio would be vacancy rich, with a lower ratio interstitial rich (8). As the crystal cools vacancies

agglomerate into octahedral voids also known as D defects or COPs. The density of the voids is about 10^6 cm^{-3} (9).

The OPE C and D were also tested on CZ substrates ((100), p-doped, thickness: 720μm) and FZ material ((100), p-doped, thickness 720μm). The wafers used were from an era where the CZ crystal growth process produced a high density of grown-in defects (D defects, COPs) .The same material was also used in a parallel comparative study in which it was etched with non-diluted Secco (3). The etch pits produced with Secco were round (figure 12, 14) while those produced with the OPE solutions were square-shaped (Figure 13, 15). Both correspond to D defects, vacancy agglomerates which form octahedral shaped cavities (voids) in the crystal lattice. In the case of the CZ material the etch pit densities found with the OPE solutions were ten times higher than those found with Secco (table III).

TABLE III. Comparison of the defect densities found after etching with Secco and the OPE D solution

material used	Etch solution used	removal (μm)	defect density cm^{-3}
Si-Bulk (CZ material)	Secco	2.68	$5.52*10^6$
Si-Bulk (CZ material)	OPE D	1.7	$5.56*10^7$
Si-Bulk (FZ material)	Secco	3	$1.178*10^6$
Si-Bulk (FZ material)	OPE D	1.7	$2.785*10^6$

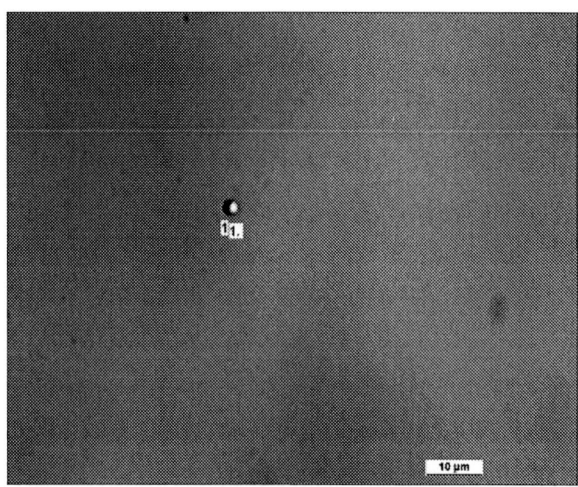

Figure 12: Optical microscope image of a round-shaped etch pit found after etching of CZ material with Secco solution.

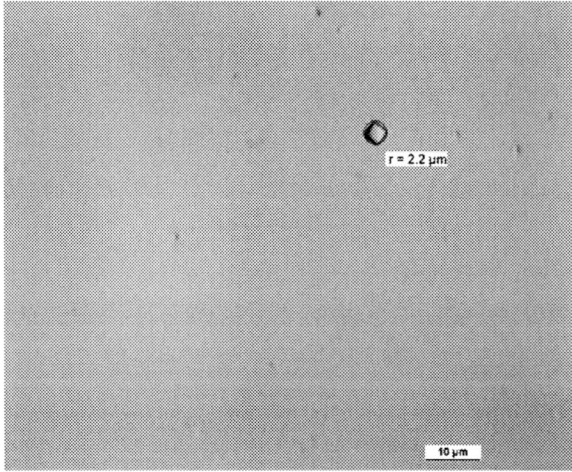

Figure 13: Optical microscope image of a square-shaped etch pit found after etching of CZ material with OPE D solution.

Figure 14: Optical microscope image of round-shaped etch pits found after etching of FZ material with Secco solution.

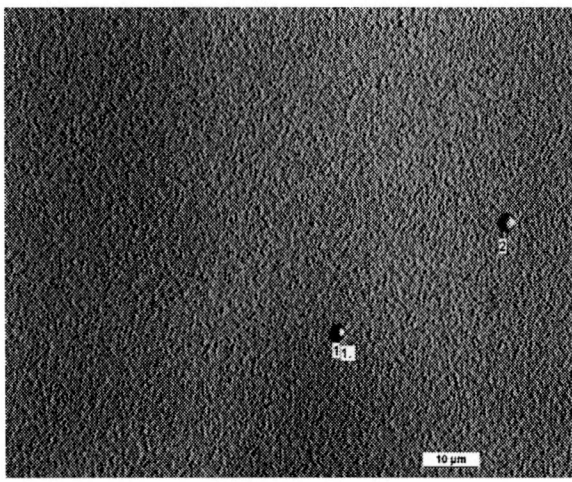

Figure 13: Optical microscope image of a square-shaped etch pits found after etching of CZ material with OPE D solution.

Summary and Conclusions:

The Organic Peracid Etches are able to reveal defects in SOI and CZ or FZ grown bulk material. In the case of the SOI material hillocks were found after etching. Detailed investigations by SEM and TEM showed a little hole in the centre of the hillock and an underlying cavity in the BOX layer.

In the case of the bulk material square-shaped etch pits were obtained which are caused by vacancy agglomerates (D defects).

The Organic Peracid Etches exhibit very low removal rates (nm/min) which are a factor of 500 lower than standard Secco etch. They provide superior capability in the delineation of microdefects: the defect densities found after etching are ten to twenty times higher than those obtained with the Secco reference.

Acknowledgments

The authors would like to thank Doris Ceglarek for SEM measurements, Martin Lommel for recording the AFM images and Yvonne Filbrandt for assistance with the manuscript.

References

1. C. Mazure and G.K. Keller, *The Electrochemical Society's Interface,* p. 34-40 (2006).
2. C. Mazure and A.J. Auberton-Herve in *Proceeding of the 35th European Solid State Device Research Conference, IEE,* p. 29 (2005).
3. F. Secco d'Aragona, *J. Electrochem. Soc., Volume 119, p. 948-951* (1972).
4. W.C. Dash, *J. Appl. Phys., 27, 1193* (1956)
5. Daniel Swern, *Chem Rev., 45, 1, (1949).*
6. European Patent Application, *Publication number: 0 452 120 A1* (1991).
7. T.Abe and S.Maruyama, Denki Kagaku 35, 149 (1967); A.J.R.deKock, *Appl.Phys.Letters 16*, p 100 (1970)).
8. R. Falster and V.V. Voronkov, *MRS Bulletin, June 2000,* p. 28-32
9. Luciano Mule`Stagno, in *Semiconductor Silicon 2002* Ed.H.R.Huff, L.Fabry and S.Kishino, *Electrochemical Society Proceedings Volume Series PV 2002-1*, p. 297-301 (2002).

CHAPTER 2

EUROPEAN PROJECT ANNA
(ANALYTICAL NETWORK FOR NANOTECH)

ECS Transactions, 10 (1) 35-49 (2007)
10.1149/1.2773974 ©The Electrochemical Society

Metrology, Analysis and Characterization in Micro- and Nanotechnologies - A European Challenge

L. Pfitzner[a], A. Nutsch[a], R. Oechsner[a], M. Pfeffer[a], E. Don[b], C. Wyon[c,d], M. Hurlebaus[e]

[a] Fraunhofer Institut für Integrierte Systeme und Bauelementetechnologie
Schottkystrasse 10, 91058 Erlangen (Germany)
[b] Semilab, Prielle Kornelia u.2, H-1117 Budapest, Hungary
[c] STMicroelectronics, 850 rue Jean Monnet, 38926 Crolles, France
[d] CEA-LETI/Minatec, 17 rue des martyrs, 38054 Grenoble, France
[e] Draeger, Revalstrasse 11, 23560 Luebeck, Germany

Europe offers excellent expertise in the area of metrology, analysis and characterization in micro- and nanotechnologies. This expertise is borne by research institutes, academia, small medium enterprises, and industry. The European approach to develop synergies, to enhance the existing technologies, and to develop innovative methods is displayed by two projects named Analytical Network for Nanotech (ANNA) and Semiconductor Equipment Assessment for NanoElectronic Technologies (SEA-NET). ANNA is an infrastructure initiative focusing on the integration of independently operating laboratories. This multi-site laboratory forms a collaborative, synergistic network of analytical working scientists and pre-existing institutions. The objective of SEA-NET is to validate emerging semiconductor manufacturing equipment, also including metrology tools for advanced process requirements for the next technology nodes. The prototype equipment assessment is performed in cooperation of tool suppliers with semiconductor manufacturers and research institutes. Results of non-contact resistivity measurements, X-ray metrology platform, and a gas detection system based on IMS technology are presented.

Introduction

For further continuation of micro- and nanoelectronics along Moore`s law, an enormous effort focusing on the understanding and controlling of material properties, dimensions and defects towards atomic level is required. Therefore, huge challenges in improved capabilities of metrology and analysis equipment, of preparatory know-how, of off-line, in-line and in-situ characterization and advanced process control have to be tackled. Metrology and instrumental analysis, already seen as a European strength, is enhanced by an impressive set of measures and collaboration. A short introduction into these focal activities will be presented.

35

ANNA – Analytical Network for Nanotech

Background

The continuous miniaturization supported by newest technologies enables advanced micro- and nanoelectronics. An enormous worldwide R&D effort focuses on the understanding and controlling of material properties and of dimensions at atomic level. Crucial for groundbreaking new developments is the availability of appropriate analytical infrastructures. New materials, and the demand of improved detection sensitivities regarding contaminants, provide huge challenges concerning the capabilities of current analysis equipment and expertise. In the past, European laboratories - with core competencies in materials characterization and trace analysis - worked mostly independently.

Objectives and Outline

ANNA focuses on the integration of independently operating laboratories (1). This multi-site laboratory forms a collaborative, synergistic network of analytical working scientists and pre-existing institutions (2-8). The base of the analytical network is the competence and expertise of the distributed laboratories sustained by intense development of new innovative technologies and methods. Complementary metrological and analytical capabilities support the integration and development process. At the same time, the offer of analytical services and transnational access enhances future research and development of the multi-site laboratory. A goal of ANNA is to provide accredited 'Golden Laboratories'. For an overview on the overall outline of ANNA see Figure 1. The following are the objectives of ANNA:

- To integrate and enhance analytical resources,
- To create a centre of excellence of analysis for nanotechnologies and a multi–site laboratory,
- Long term vision: integrated distributed laboratory.

The approach to concentrate research resources is supported by the instrument of an Integrated Infrastructure Initiative (I3) structure supported by European Commission Contract ANNA RII3 026134. An objective is to offer the research community Transnational Access to infrastructures. For formation and integration of the multi-site laboratory, ANNA uses three instruments; the implementation is described in detail in the following paragraphs:

- Networking,
- Transnational Access,
- Joint research.

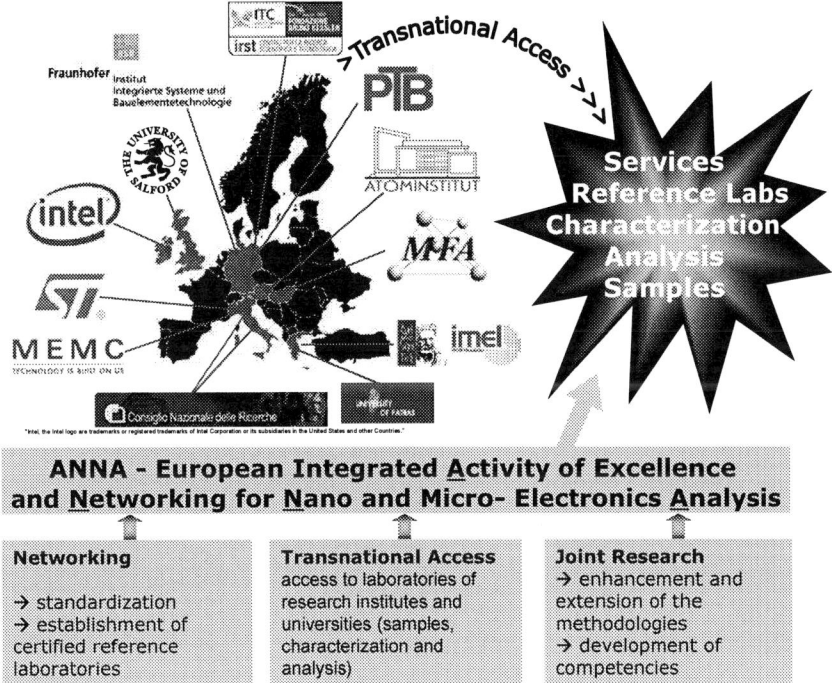

Figure 1. Overview on the outline of ANNA, the project components and the offered services available to researchers, small medium enterprises, and industry.

Networking:

The underpinning component of ANNA is the networking activity. The networking activities for ANNA have the aim to establish 'Golden Laboratories', to integrate the multi-site laboratory, and to standardize samples and methods. Furthermore, ANNA initiates a coherent network of scientists and the service of accredited test and calibration laboratories. ANNA comprises the following specific activities for networking:

Establishment of analytical reference laboratories: Researchers and industrial users require comparable results when analytical methodologies are supplied by different research institutions. Moreover, industrial partners and suppliers are usually certified according to ISO/TS 16949. Therefore, the research institutes supplying services and research to industry apply for accreditation in order to fulfill the industrial certification requirements. The core competencies of the project partners will be used to establish reference laboratories for analytical methodologies named 'Golden Laboratories'. The analytical techniques will be accredited according to ISO 17025:2005 or certified according to ISO 9001:2002.

Formation of a multi-site laboratory: During the process of forming the multi-site laboratory, the 'Golden Laboratories' of the project partners are joined. The multi-site laboratory offers a wide range of analytical services. An overview on the analytical techniques available within ANNA is given in Table I. Samples offered by the laboratory are for example test structures, wafers, and contamination and calibration standards. The integration of the multi-site laboratory comprises the development of electronic integration environment by an internet interface. This is supported by the supply of user documentation and manuals for the infrastructures. In the future, the internet interface will provide on-line sample and result tracking.

TABLE I. Overview Analysis / Metrology / Characterization.

Classification	Methods
X-Ray technologies	TXRF, TXRF - NEXAFS, GIXRF, XRR, XRD
e-beam technologies	TEM, HRTEM, STEM
Ion beam technologies	SIMS, ToF-SIMS, MEIS
Surface characterization	XPS, AES, UPS, LEISS, EELS
Chemical analysis	AAS, GCMS, Sample Preparation
Electrical characterization	C-V, C-T, I-t, C-G, SPV, DLTS
Optical metrology	spectroscopic ellipsometry, defect inspection, FTIR

Standardization of samples and methodologies: Calibrated analysis and validated results from characterization for samples of micro- and nanotechnologies is currently rarely found at a European level. Standard and reference samples according to industrial and researchers' requirements offered by the partners of ANNA will enable comparison among complementary analytical techniques and round robin tests enhancing the analytical capabilities. Matching and benchmarking of similar analytical techniques at different installations and round robin tests is a key point to sustain the progress towards standardized and calibrated results from the integrated laboratory and each of the project partners. Furthermore, the offer to external labs to compare and match their analytical capabilities with the reference laboratory enhances the analytical capabilities.

Transnational Access:

Transnational Access to 18 infrastructures (laboratories, clean room, and metrology) at 8 locations is offered through ANNA. An overview on the infrastructures available to users is given in Figure 2. Transnational access to ANNA instrumentation and analytical services is available either in person 'hands-on' or remotely by 'electronic communications'. Transnational access enhances and improves the performance and analytical capabilities of the multi-site laboratory.

Joint Research:

This instrument has the objective to perform joint research into higher performance techniques, instrumentation or technologies to improve the service provided by the multi-site laboratory. The optimized approach to any 'borderline' metrology problem typically requires the application of supplementary analytical techniques. Some of these are not commonly available to industry, especially in situations where techniques, analytical methodologies and individual expertise are pushed to the limit in dealing with critical issues faced by industrial partners. At the same time, the availability of complementary competences is crucial for the advancement of any analytical methodology through cross-comparison, round-robin, and benchmarking of results.

PTB	TA1: PTB @ BESSY II • Synchrotron radiation beam lines for TXRF, GIXRF, XRR	TA5: MFA - facility • Ellipsometry • Makyoh	MFA
ITC irst	TA2: irst - SIMS & MICRO • SIMS • ToF-SIMS IV • SEM JSM 7401F • AFM, XPS	TA6: IISB - laboratories • Ultra trace analysis • Organic contamination analysis • Wafer surface preparation and contamination	IISB
	TA3: CNR - STEM facility • STEM	TA7: IMEL - laboratories • Electrical and optical characterisation • Fabrication of test structures	imel
ATOMINSTITUT	TA4: Atominstitut • ATI-x-ray lab	TA8: MEIS – facility at Daresbury laboratory • MEIS	The University of Salford

Figure 2. Overview on the Infrastructures available during Transnational Access through ANNA.

A vital aspect and overall target of the ANNA project is the synergy in the ensemble of the joint research activities and the project partners. Furthermore, the collaboration between leading edge micro- and nanoelectronic industries and research institutes with specific core competences in those fields of metrology becomes more and more critical for future technology. This collaboration is essential for the joint research. Several analytical challenges, related with Ultra Large Scale Manufacturing (ULSI) silicon-based device manufacturing, are being confronted in the joint research of ANNA. Therefore, the research of ANNA covers a relatively broad spectrum of crucial topics related to metrology in silicon-based device manufacturing. They range from surface-oriented issues (surface contamination detection and quantification), ultra-thin film and shallow implants characterization as well as bulk defect and strain investigations. The research activities are highlighted in the following paragraphs.

Highly Sensitive Detection of Inorganic Contamination from Li to U: The detection of inorganic contaminants in microelectronics is increasingly challenging due to the requirements arising from the introduction of new technologies. A target is the development of new methodologies and metrologies for the detection of low concentration contaminants in silicon and in new materials, and the evaluation of their impact on the electrical properties of the materials and devices produced. This includes the following specific research topics:

- Understanding of the electrical properties of different contaminants and their impact on the device performances for the introduction of novel production technologies,
- Development of novel methods for new requirements (analysis of SOI, fast diffusers),
- Improvement of detection limits and analysis of light elements by TXRF.

Comprehension of Organic Contamination on Wafer Surfaces: It is the objective to study the role and impact of organic compounds on wafer surfaces relevant to nanotechnologies and advanced microelectronics. Organic wafer surface contamination and comparison of different detection methodologies are studied in detail. At the end of the JRA a deeper comprehension of organic contamination and enhanced detection methodologies are expected.

- Comprehension of surface mechanisms of organic contamination,
- Correlation of different methodologies/benchmarking,
- Impact determination of specific contamination on the manufacturing process and e.g. reliability.

Development of Ultra-Shallow Junction Depth Profiling: The objective is to define methodologies able to give a more complete characterization of ultra-shallow dopant distributions, in particular for Ultra-Shallow Junctions (USJ) applications, on either Si or advanced substrate or new materials, as demanded by the present and future microelectronics technological nodes. These methodologies will be the result of the comparison between the several complementary techniques (SIMS, MEIS, GIXRF, and STEM) available within the integrated infrastructure.

- Exploring and understanding the complementariness of the techniques available within the consortium for ultra-shallow distributions characterization in Si. Definition of state-of-art methodologies for SIMS for ultra-shallow depth profiling in Si comparing results with the ones coming from other techniques,
- Development and application of XRF or STEM techniques to the characterization of ultra-shallow distributions,
- Extension and application of the previous results to new substrate materials as required by the future technology needs: SiGe, Ge, strained-Si, silicon on insulator,
- Creation of a set of standards or fully characterized samples in order to assess the quantitative results of the different techniques.

Nanofilms Characterization: The objective is to develop methods for the chemical, structural, optical and electrical characterization of films of nanometer thickness through the application of a range of high resolution techniques for the analysis of 1-10 nm thick (oxy)nitrides and high k materials. The achievement of optimized depth resolution, good sensitivity and reliable quantification are the fundamental issues in this research activity. The techniques involved include HREM, GI-XRF, TOF-SIMS, MEIS, XPS, UPS, AES, LEIS, ellipsometry and electrical characterization.

- Improvement of measurement techniques for the accurate determination of the thickness of thin layers and to yield reliable compositional analysis,
- Development of protocols and analysis techniques for thin films for example XRR and/or TXRF and XRF, GIXRF, surface spectroscopic technique,
- Development and standardization of electrical characterization techniques for nm thin film dielectrics,
- Study on selected new high-k materials applying methodologies and results obtained from oxynitrides studies.

Crystal Defects and Strain in Device Processing: Mechanical stress is an important parameter in present silicon devices, because it can both, enhance device performances and destroy device functionality by inducing crystal defect formation. Therefore, the mechanical stress should be determined with the required sensitivity at the device size scale. This activity aims at setting up a method to measure the mechanical stress in present and future generation devices by TEM-CBED. The results of this technique will be validated by comparison with electrical measurements of stress-sensitive devices and with the results of numerical calculations.

- To set up a protocol to quantitatively determine the strain tensor in silicon nanodevices from STEM/CBED patterns in the most automatic way with the best spatial resolution,
- To check the validity of the results obtained by this method by comparison with numerical calculation of the mechanical stress in the device process flow,
- To identify the critical strain values significant for device operation, namely the critical strain values to have
  - a significant impact on carrier mobility, hence of the electrical characteristics of transistors,
  - crystal defect formation and hence the destruction of device functionality,
- To identify the impact of substrate properties (i.e. defects like oxygen precipitates or ad hoc fabricated strained silicon films) on the mechanical strain in silicon, and assessment of the ability of the TEM/CBED technique to measure such strain,
- Strain detection obtained from the deformation (curvature) of the wafer surface measured by Makyoh topography, an optical metrology tool.

Characterization of Nanocrystals: Nanocrystalline semiconductors embedded in dielectric matrices (e.g. silicon rich oxide, ion-implanted silicon oxide, SixOyNz, etc.) are currently under investigation for use in Si-photonics and in memory devices. The aim is to develop and improve metrologies for the measurement of nanocrystal properties.

- Production of Si nanocrystals in a controlled way using different growth techniques,
- Benchmarking, assessment, development, and improvement of metrology of techniques for nanocrystal characterization,
- Better understanding of the correlation between device properties, structure, and preparation.

SEA-NET – Semiconductor Equipment Assessment for NanoElectronic Technologies

An example for assessment of metrology equipment and driving new research is included in SEA-NET (9). The main objective of the project partners of SEA-NET is to validate emerging semiconductor manufacturing equipment and metrology equipment for future technology generations at advanced user sites. The project stimulates the European approach to initiate sustaining partnerships amongst equipment industry, IC industry and research institutions, addressing needs of the next 2 to 10 years.

Introduction

Back in the 60ies and early 70ies of the last century, laboratory equipment was built and used by the IC manufacturers to produce the first series of integrated circuits. Those tasks were taken over by a rapidly growing number of equipment manufacturers, who now built the tools according to the specifications of the semiconductor manufacturer. Today we see a completely different scenario: the equipment has to be delivered with a mature and production-proven process, the development of process technology, and even part of the process integration has to be provided by the equipment industry. However, innovative processes and new metrology methods are often invented and developed by academia or small and medium enterprises, who are lacking an infrastructure necessary for the maturization of according tools. This has led to new collaboration and networking needs amongst research, equipment manufacturing and IC manufacturing. Within a challenging European research project, a novel strategy of integration and collaboration of several equipment manufacturers, research institutions, and IC manufacturers for advanced equipment evaluation and long termed equipment research was established, the 'Integrated Research Project SEA-NET'.

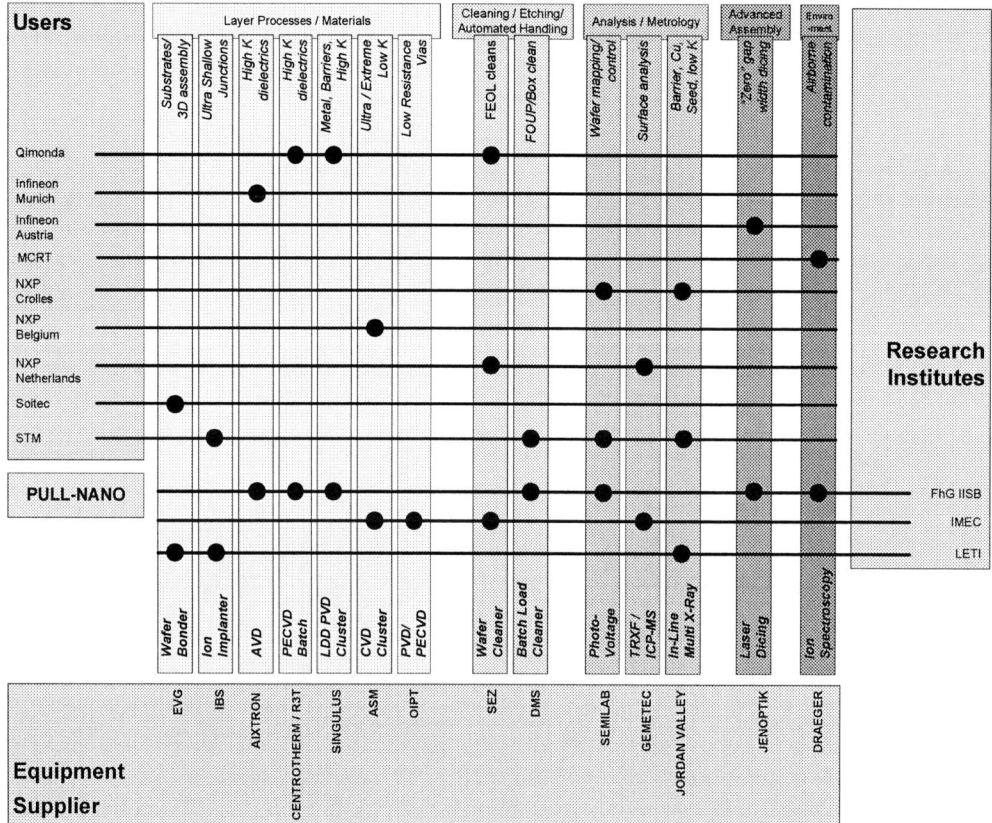

Figure 3. Structure of the Integrated Project SEA-NET funded by the EC

Strategic Targets of SEA-NET

Proven equipment, key-process advances, and equipment control introduced in a cost-effective manner now dictate the timing for the introduction of advanced technology nodes and new device architectures, rather than sequential lithographic scaling. SEA-NET is an innovative approach to speed up the development of new and advanced semiconductor manufacturing equipment and metrology tools, and to close the gap between equipment development and its integration into semiconductor production.

The main themes include: Advanced Substrate Materials, Dopant/Layer Control, Dielectric Isolation and Gate Structures, Barriers/Multilayer Interconnect and methodologies that can impact the emergence of broader functionality integration using System on Chip (SoC) or System in Package (SiP). In addition, key metrology and analytic equipment are evaluated as to their capability to accurately measure the key parameters of these layers/multilayers, particularly within automated production lines. Automated handling and cleaning equipment is featured as well, as it is essential if resulting wafers are to be transferred and evaluated within non-contaminated environments. An objective of the programme is to define a SEA-NET specific certification procedure for APC capabilities, and to shorten yield-learning cycles by predicting the yield of the currently processed wafers. This includes: model-based control, process and equipment simulation, enhanced wafer and equipment logistics, definition of advanced communication as well as man-machine-interfaces, and especially virtual equipment engineering.

SEA-NET consists of 16 sub-projects. Three of them are covering semiconductor equipment assessment for metrology tools. An overview of the objectives, including examples and results from metrology, analysis and characterization of these sub-projects are presented in Figures 3 to 6.

Subproject SP11: Lead It

Wafer mapping and process control metrology of ion implant dose has always been a key demand in semiconductor manufacturing. In the future, implant users will be required to measure and thus control - with much better than 0.1% resolution in dose value - the low dose and very low energy implants required for the nano-scale CMOS process 90, 65, 45 and 32nm technology nodes. Currently no proven and accepted equipment is available to non-destructively measure sheet resistivity of the Ultra-Shallow Junctions required for the future process nodes.

Semilab has utilized knowledge gained in developing and deploying its current family of semiconductor metrology tools to develop a new surface photo voltage (SPV) based method for fast high resolution sheet resistivity mapping to provide a total integrated solution for low dose implant monitoring. These new technologies have been proven in a prototype lab environment but within this sub-project the full measurement solution will be adapted and integrated into a SEMI standards compliant and 300mm capable automation system, so that it can be optimized within a 300mm wafer fabrication environment. The final optimization of the measurement functionality and its associated wafer processing recipes can only be achieved within a nano-scale wafer fabrication environment using state of the art USJ implanted wafers. The development of the automation system is already available, and does not require user demonstration since Semilab has already installed similar automation systems in production environments but for different applications and measurement technology.

The new measurement units will be based upon Semilab's patented probe technologies, but will also incorporate further improvements that may be patented during the subproject. The speed, spatial resolution, and dose range of the measurements will fulfill volume production requirements for implant monitoring in the 90, 65, 45 and 32nm technology nodes.

<u>Major Achievements of SP11:</u> The mechanical construction of the 300mm clean JPV sensor automation platform, including a fast wafer R-θ scanner, is completed and also a re-design of the platform to a modular concept. Test wafers supplied by potential users and R&D partners were measured and identified implanter issues not seen by other metrology tools (15). The system was installed in a 300mm wafer fab to evaluate the JPV measurement technology. The results obtained in the production environment confirmed the results previously obtained by the tool supplier. Extensive map correlation studies between JPV, Hg-4PP and conventional 4PP had been undertaken. The mean value correlation of the sheet resistance to other methods is excellent over a wide range of dose and energy. Algorithms were developed to correct for some distortion in the map due to JPV edge effects (16).

Benefits of LEAD-IT

Fully automated 300mm metrology tool for the measurement of sheet resistivity

- Non-contact metrology
- Non-destructive
- High speed
- High resolution

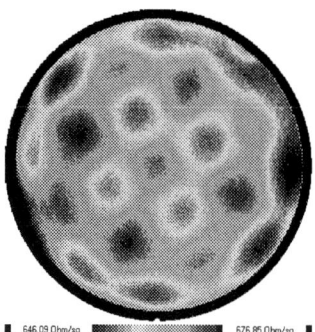

Figure 4. Results from a non-destructive high speed, high repeatability full wafer mapping sheet resistivity measurement using junction photo-voltage.

<u>Subproject SP13: MUXT</u>

Yield and reliability of advanced silicon devices are governed by Copper and low κ interconnects. Electro and stress-migration properties of copper interconnects are highly sensitive to the structural properties of Cu layers, which are influenced by the underneath barrier thin films and Cu seed layer, as well as the impurities of the electrochemical Cu deposition solutions. X-ray diffraction (XRD) could be the metrology technique of choice for monitoring the structural properties of Cu and Cu-related layers, as soon as X-ray diffraction can be fast, fully automated and can exhibit a small X-ray spot size to perform in-line monitoring on 300mm product wafers.

Porous low κ dielectrics will be introduced in ≤ 65nm generation advanced silicon devices. The in-line monitoring of the overall porosity and the pore size distribution (PSD) of low κ films is compulsory for mastering the dielectric and mechanical properties of low κ dielectrics. GI-SAXS (Grazing Incidence-Small Angle X-Ray Scattering) is the only characterization technique which could be used as an in-line metrology technique for PSD monitoring since it is viable and non-destructive. Nevertheless many obstacles remain before the use of GI-SAXS as an in-line metrology technique: decoupling the surface roughness signal from the pores scattering signal, careful calibration and/or modelling of the GI-SAXS spectra and intensity.

X-ray reflectivity (XRR) has already demonstrated its capabilities for the in-line monitoring of thickness and density of thin metallic and dielectric films. XRR will be obviously used to monitor the overall porosity of low κ dielectric layers. XRR will also be used and improved for the in-line monitoring of metallic film thickness and density of copper diffusion thin films, as their thickness will be considerably reduced from 90 to 32nm generation technologies.

The objectives of the MUXT sub-project are the development, the assessment and the improvement of a new fully automated X-ray metrology platform: 3μX, which will allow the whole monitoring of Cu/low κ interconnects by gathering on the same platform:

- Fast μ-spot X-ray diffraction technique for the in-line monitoring of the structural properties of poly-crystalline thin films (Ta/TaN, WCN...) and Cu layers on 300mm monitor and product wafers,

- GI-SAXS and XRR for the in-line monitoring of the overall porosity and PSD of low κ dielectric layers,

- XRR for the in-line monitoring of thin and thick metallic films thickness and density.

The 3μX tool will be fully assessed and improved in terms of hardware, metrology key parameters, and measurement accuracy for fulfilling the metrology requirements of 65, 45 and 32nm Cu/Low κ interconnects.

Main efforts will be devoted to the development of the fast μ-XRD equipment as well as the careful calibration or modeling of GI-SAXS spectra of low κ layers exhibiting an average pore diameter lower than 1nm.

Major Achievements of SP13: The GIX tool gathering XRR and GI-SAXS techniques was installed in a 300 mm clean-room. A site acceptance test was performed which revealed that the values of capabilities comply with 65nm process control requirements. The μ-XRD test bench developed by the tool supplier was available at its premises. This bench is used for the metrology of Ta, TiN and Cu thin films. The GIX tool was successfully used for the monitoring of Cu/low κ interconnect, Cu capping layers and MIM capacitor processes on monitor wafers. The tasks dedicated to the integration of the μ-XRD module were started.

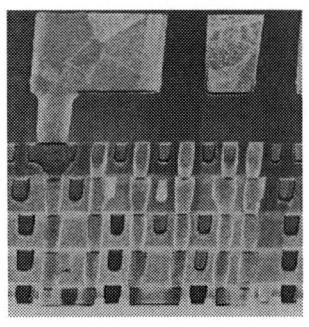

Benefits of XXR and XRD

- Fast and fully auto-mated metrology techniques for thick-ness, density (poros-ity) and texture moni-toring of thick and thin:
  - Metallic layers: Cu, TaN, TiN, ...
  - Dielectric films: high κ, low κ, ...

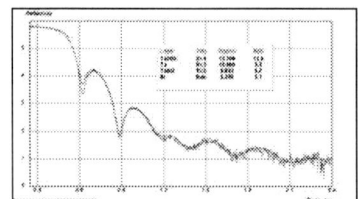

Figure 5. Results from XRR spectrum of a 13nm thick PVD Ta film deposited on Si

Subproject SP17: Mosaic

The main objective addressed in this sub-project is to specify, design, and evaluate a monitoring system for airborne molecular contamination (AMC) applicable for process control and yield enhancement tasks in semiconductor manufacturing areas and micro-environments. The most important technical objectives to be achieved are:

- In-situ monitoring of acidic gas contamination (Acid AMC monitor),

- In-situ monitoring of basic gas contamination (Base AMC monitor),

- In-situ monitoring of certain organic contamination species (VOC AMC monitor).

Thus, this sub-project should achieve the design and evaluation of an airborne mo-lecular contamination system with optimized response times and detection sensitivities for in-situ applications in current and future semiconductor manufacturing processes with technology nodes even below 65nm. Strategic objective of the sub-project is also to strengthen the competitiveness of the European semiconductor equipment manufacturer and semiconductor manufacturer community, as the current metrology market is domi-nated from non-European companies.

The basis for the AMC monitoring system is an Ion Mobility Spectroscopy platform of Draeger Safety. Detailed assessment by the sub-project consortium including a leading semiconductor manufacturer, a design and manufacturing company for cleanrooms and micro-environments and one of the leading European research institutions will enhance the in-situ monitoring system device in terms of response times and sensitivity to meet the specifications of the semiconductor manufacturers according to the industry roadmaps for future technology nodes below 65nm.

The sub-project comprises of three main sub-project phases:

- Prototyping of IMS devices according to customer requirements and technical speci-fications from the ITRS and SEMI standards,

- System evaluation at research partner including benchmarking tests with comparable analytical instruments,

- System evaluation at industrial user sites.

Appropriate sub-project management, dissemination and communication structures will be implemented according to the size and complexity of the sub-project.

Major Achievements of SP17: A gas detection system based on IMS technology was designed without using membranes in the gas inlet system. Target gas detection and the discrimination from other gases take place in a beta drift tube. As one main improvement of the AMC monitor, the Tritium source has been re-designed and its beta radiation activity is 0.3 GBq (compared to 0.5 - 4.4 GBq for the previous system). This activity is small enough to meet the requirements of industrial users for handling, transportation and operation.

Algorithms for signal processing have been developed and the software implementation for signal processing algorithms for the volatile organics (VOC), base and acid IMS versions were performed. The metrological concept for the VOC and base IMS version were implemented in prototype systems. The VOC IMS prototype has been tested in comparison to off-line analyses showing the required response times and single ppb detection limits (response curve to NMP see Figure 6). Currently site validation tests for the VOC IMS are conducted at a major European semiconductor manufacturer. The base IMS prototype is now in the benchmarking test before it will be tested at a semiconductor manufacturer site. The acid IMS prototype is being integrated at the tool manufacturer site.

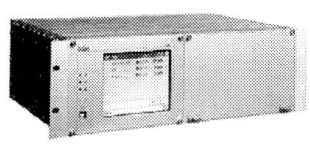

Benefits of Mobile IMS

- Improved Detection Limits (ppb)
- Improved Reliability
- Direct AMC Detection
- On-Line Detection

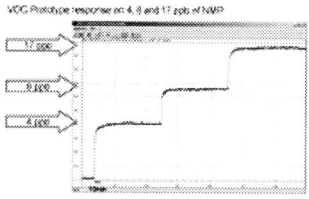

Figure 6. Results from VOC prototype of 4, 8 and 17ppb of N-Methyl-2-pyrrolidon (NMP)

Summary and Conclusion

The challenges of further advances in metrology, analysis and characterization in micro- and nanotechnologies require a paradigm shift in collaboration. Above, new approaches of collaboration amongst universities, research institutions, analytical laboratories, suppliers, and users of measurement and analytical tools were shown. Such new approaches were identified as an ideal way to create a better academic and industrial infrastructure, including interests from and benefits for small and medium enterprises (SME). Above, the collaboration with Networking, Transnational Access, and Joint Research towards analytical and metrology laboratories as well as improvements for an improved tool development and assessment was described in detail. Organization and structures, methodologies, and results were given.

Well recognized is the need for a global collaboration. Interested partners from academia as well as from industry are encouraged to benefit from the above described activities (1, 9) on a global scale. Also a new global program called IMS - INTELLIGENT MANUFACTURING SYSTEMS – is considered to become a world wide foundation for further enhancements of metrology and analytics (13). This IMS initiative encourages the formation of international research consortia to address 21st century manufacturing challenges. The scheme aims to boost also metrology and analytical instrumentation, and its integration into advanced manufacturing control and standardization issues. IMS members currently include the EU, Norway, Korea, Japan, Australia, Switzerland, Canada, and the United States of America (14). Recent consultations revealed the importance of the control of manufacturing processes as well as the improvements of metrics and metrology in general, but also especially for micro- and nanotechnologies including new materials and materials systems – touching a European strength and a global challenge.

Acknowledgments

The authors would like to thank the European Commission for partly funding the projects ANNA (contract no. 026134-RII3) and SEA-NET (contract no. 027982) within the 6th Framework Program. The authors also would like to acknowledge numerous colleagues at the partner sites for their valuable contribution.

References

1. Project website ANNA: www.anna-i3.org
2. B .Beckhoff, 'Wafer contamination analysis, speciation and reference-free nano-layer characterization using synchrotron-based X-ray spectrometry', ECS Satellite Symposium 'Analytical Techniques for Semiconductor Materials and Process Characterization V' ALTECH 2007 joint with ESSDERC, Munich, Germany, (2007); published in this volume: ECS Transactions – ALTECH 2007
3. A. Armigliato, 'Structural and Analytical Characterization by Scanning Transmission Electron Microscopy of Silicon-based Nanostructures', ECS Satellite Symposium 'Analytical Techniques for Semiconductor Materials and Process Characterization V' ALTECH 2007 joint with ESSDERC, Munich, Germany, (2007); published in this volume: ECS Transactions – ALTECH 2007
4. P. Petrik, 'Ellipsometric characterisation of semiconductor nanocrystals, thin films, and surface', ECS Satellite Symposium 'Analytical Techniques for Semiconductor Materials and Process Characterization V' ALTECH 2007 joint with ESSDERC, Munich, Germany, (2007); published in this volume: ECS Transactions – ALTECH 2007
5. S. Ladas, 'Thin film analysis and model interface characterization studies of relevance to microelectronics', ECS Satellite Symposium 'Analytical Techniques for Semiconductor Materials and Process Characterization V' ALTECH 2007 joint with ESSDERC, Munich, Germany, (2007); published in this volume: ECS Transactions – ALTECH 2007
6. J. A. Van Den Berg, 'Medium energy ion scattering (MEIS) for the characterisation of ultra shallow implants and high-k dielectric films', ECS Satellite Symposium 'Analytical Techniques for Semiconductor Materials and Process Characterization V' ALTECH 2007 joint with ESSDERC, Munich, Germany, (2007); published in this volume: ECS Transactions – ALTECH 2007

7. G. Borionetti, 'Surface Microdefects control during Chemical Mechanical Polishing of Silicon wafers: an example of in line manufacturing process control', ECS Satellite Symposium 'Analytical Techniques for Semiconductor Materials and Process Characterization V' ALTECH 2007 joint with ESSDERC, Munich, Germany, (2007); published in this volume: ECS Transactions – ALTECH 2007

8. D.Codegoni, M.L. Polignano, 'Molybdenum contamination in indium implantation', ECS Satellite Symposium 'Analytical Techniques for Semiconductor Materials and Process Characterization V' ALTECH 2007 joint with ESSDERC, Munich, Germany, (2007); published in this volume: ECS Transactions – ALTECH 2007

9. Project website SEA-NET: www.sea-net.info

10. D. Warren and J. M. Woodall, in *Semiconductor Cleaning Technology/1989*, J. Ruzyllo and R. E. White, Editors, PV 90-9, p. 371, The Electrochemical Society Proceedings Series, Pennington, NJ (1990).

11. F. P. Fehlner, *Low Temperature Oxidation: The Role of Vitrous Oxides*, p. 23, Wiley Interscience, New York (1986).

12. N. J. DiNardo, in *Metallized Plastics 1*, K. L. Mittal and J. R. Susko, Editors, p. 137, Plenum Press, New York (1989)

13. Program website: www.ims.org

14. Organization website: www.mel.nist.gov/ims/docs/ims_org.pdf

15. E. Don et al, IIT 2006 11th -16th June 2006. AIP Conf. Proc.Vol.886 p.534 , 2006

16. E. Don et al in Proc. INSIGHT-2007 May 6-9, 2007, Napa, California. USA

50

Wafer Contamination Analysis, Speciation and Reference-free Nanolayer Characterization using Synchrotron Radiation based X-ray Spectrometry

B. Beckhoff[*], R. Fliegauf, M. Kolbe, M. Müller, B. Pollakowski, J. Weser, and G. Ulm

Physikalisch-Technische Bundesanstalt, Abbestraße 2-12, 10587 Berlin, Germany

Total-reflection, grazing-incidence and reference-free X-ray fluorescence analysis employing synchrotron radiation can substantially contribute to the contamination analysis and speciation on semiconductor surfaces as well as to the non-destructive characterization of nanolayered systems allowing for high information depths.

Semiconductor characterization by reference-free X-ray spectrometry

The continuing development of analytical methods based on X-ray spectrometry and related instrumentation by the Physikalisch-Technische Bundesanstalt (PTB), Germany's national metrology institute, is dedicated to high-end investigations in the R&D of semiconductor samples related to industrial applications. The use of monochromatized synchrotron radiation at the 1.7 GeV electron storage ring BESSY II has led to developments in the non-destructive investigation of wafer surface contamination, speciation and nanolayered materials by X-ray spectrometry. X-ray fluorescence spectra of various semiconductor samples can be recorded in different geometries with respect to the sample surface: varying the incident angle of the synchrotron radiation from close to zero degrees, i.e. total-reflection geometry (T), over grazing incidence (GI) as well as conventional X-ray fluorescence (XRF). In TXRF geometry, only the surface is analyzed, whereas GIXRF and XRF provide information from subsurface layers respectively the bulk of a sample.

For analytical investigations relevant for the semiconductor industry, the PTB can handle 200 mm and 300 mm silicon wafers as well as smaller semiconductor wafers in its TXRF, GIXRF and XRF instrumentation (1,2). To prevent undesired cross-contamination of the wafers, the instrumentations are protected by mobile cleanrooms (cfr. fig. 1). The instrumentation for 200 mm and 300 mm Si wafers is equipped with an EFEM module to which SMIF and FOUP boxes can be directly attached. PTB's XRF instrumentation is equipped with calibrated photodiodes and energy-dispersive detectors (3), allowing for the absolute determination of the incident radiant power and detected fluorescence count rates. Routine analysis of wafer surfaces and nanolayers involves XRF spectra deconvolution based on both detector response functions and physical modeling of background components such as resonant Raman scattering (4) at the wafer substrate. Furthermore, *reference-free quantitation* of surface contamination as well as of the thickness and composition of nanolayers is ensured by the knowledge of all relevant parameters: The spectral efficiency of detectors employed, the solid angle of detection

---

[*] corresponding author; email: Burkhard.Beckhoff@PTB.de

involving the beam profile of the excitation radiation, and the fundamental atomic data, such as fluorescence yields and absorption cross-sections of the elements of interest.

Figure 1. TXRF, GIXRF and XRF instrumentation **a)** for 15 mm through 75 mm Si or SiC wafers and **b)** for 200 mm and 300 mm Si wafers located in the focal plane of the PTB plane-grating monochromator (PGM) beamline for undulator radiation at the electron storage ring BESSY II.

Contamination control on semiconductor surfaces

The use of undulator radiation in the PTB laboratory at BESSY II for its XRF instrumentation is advantageous for contamination control on semiconductor surfaces as it provides very high photon fluxes of polarized radiation in the soft X-ray range for the efficient excitation of *light elements* in TXRF analysis. Optimizing the respective excitation conditions such as the angle of incidence and the incident photon energy, detection limits of light elements such as C, N, F, Na, Mg and Al in the fg range can be achieved. The absolute calibration of the instrumentation allows for the determination of wafer surface contamination by *reference-free analysis*. Figures 2 and 3 show TXRF spectra recorded from the surface of 200 mm Si wafers involving contamination by both light elements and transition metals. The spectra are deconvoluted with respect to the fluorescence lines of the contamination elements and physical background components such as bremsstrahlung, resonant Raman and Rayleigh scattering. The fluorescence intensities deduced are used to calculate the respective elemental mass depositions taking into account the incident radiant power, the efficiency of the Si(Li) detector employed and the effective solid angle of detection without using any reference sample.

Due to the energetic tunability of synchrotron radiation, near-edge X-ray absorption fine structure (NEXAFS) can be combined with TXRF analysis to contribute to the *speciation* of contamination by light or organic compounds at extremely low levels (5). *Nanoparticles* deposited on wafer surfaces can be likewise analysed and speciated (6), whereby angular scans can be employed to distinguish between particle-, layer- and bulk-type contamination.

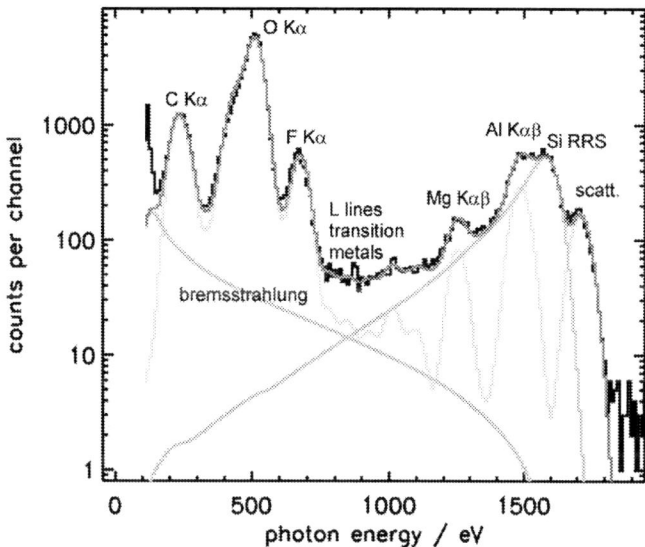

Figure 2. TXRF spectrum deconvoluted with experimentally obtained detector response functions with respect to C, O, F, Mg and Al-K fluorescence as well as transition metal L fluorescence radiation, bremsstrahlung background, resonant Raman scattering (RRS) background and Rayleigh scattering (scatt.) of the excitation radiation at 1713 eV.

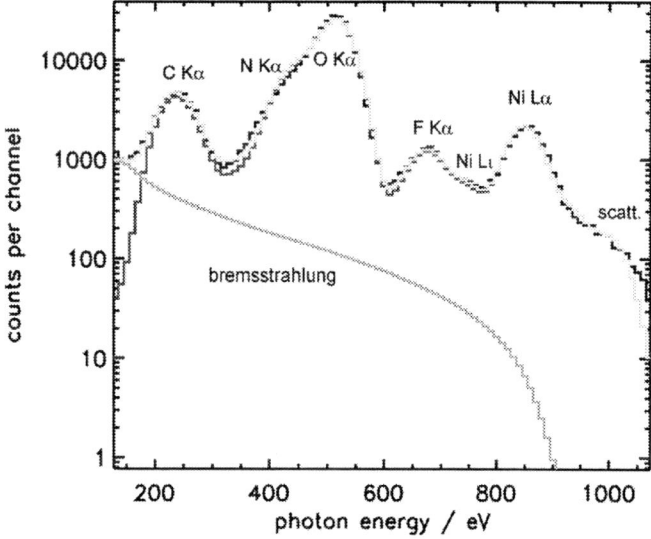

Figure 3. TXRF spectrum deconvoluted with experimentally obtained detector response functions with respect to C, N, O, and F-K fluorescence as well as Ni-L fluorescence radiation, bremsstrahlung background and Rayleigh scattering (scatt.) of the excitation radiation at 1000 eV.

Reference-free nanolayer characterization

For the R&D of semiconductor devices, it is not only important to minimize surface contamination, but also to determine nanolayer characteristics. Reference-free XRF can provide information on both the *nanolayer thickness* and *composition,* even when no reference samples for new materials are available (7). The layer thickness obtained for transition metals as well as silicon dioxide layers by XRF is well in line with the thicknesses determined by X-ray reflectometry (XRR) within their respective uncertainties. XRR also offers information on the layer density.

Figure 4 shows the spectra of two transition metal layers deposited on Si wafers recorded in a conventional orthogonal XRF beam geometry. The Ni and Cu layer thicknesses determined by reference-free XRF quantitation of the K fluorescence radiation (20.2 nm ± 1.2 nm, respectively 20.4 nm ± 1.2 nm) are in line with the thicknesses obtained by complementary XRR investigations (20.1 nm ± 0.2 nm, respectively 20.0 nm ± 0.2 nm). When dealing with more complex nanolayered systems, e.g. involving two different transition metal layers, the sequence of these layers can be determined by analyzing the L fluorescence of both elements due to more drastic absorption effects in the soft X-ray range.

Figure 5 shows XRF spectra of both a native and a thermal SiO_2 nanolayer on Si recorded in the soft X-ray range. The spectra deconvolution allows for the determination of the respective SiO_2 layer thicknesses and the top C contamination layer without employing any well-known reference layer.

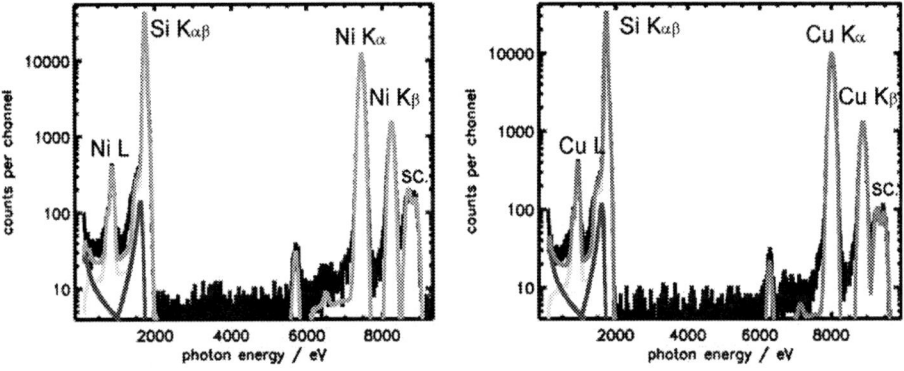

Figure 4. X-ray fluorescence spectra of two different transition metal nanolayers on Si deconvoluted with detector response functions. In the soft X-ray range below 1800 eV a continuous bremsstrahlung background and a resonant Raman scattering background induced by Si-K fluorescence radiation are depicted. The Compton and Rayleigh scattered (sc.) excitation radiation is also indicated. The figure to the left shows a Ni layer, whereas the figure to the right shows a Cu layer. The black lines indicate the measured spectra. The different gray lines are the convolution of the fluorescence line energies with the detector response functions, their fit to the spectrum to derive the net intensities for the fluorescence lines, and the different background components.

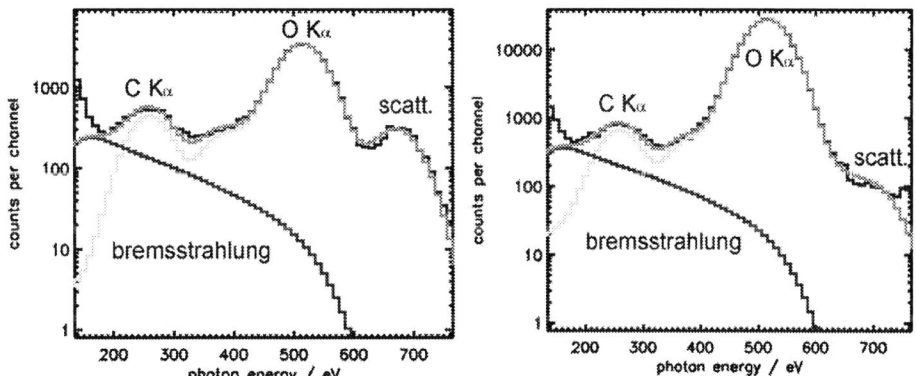

Figure 5. Soft X-ray fluorescence spectra of two different SiO$_2$ nanolayers on Si deconvoluted with detector response functions assuming a continuous bremsstrahlung background. The scattered excitation radiation is also indicated. The figure to the left shows a 1 nm thick native oxide layer whereas the figure to the right shows a thermal oxide of 7 nm thickness. A 0.5 nm thick C contamination layer was found on top of both SiO$_2$ nanolayers by reference-free XRF quantitation.

For the speciation of the nanolayer composition, grazing incidence XRF allows for different penetration depths by varying the incident angle. At a given incident angle, GIXRF can be combined with NEXAFS, revealing information on the depth profile of the chemical layer structure. Due to the higher information depths of photons as opposed to electrons, *elemental depth profiling* by GIXRF allows for the analysis and speciation of buried nanolayers from only a few nm to several hundreds of nm below the surface as well as of nanolayers on top of the substrate surface.

Summary

Reference-free methods in X-ray fluorescence analysis can be effectively used to gain information concerning semiconductor surface contamination and nanolayered systems. When performing total-reflection XRF combined with synchrotron radiation, lowest detection limits can be achieved for off-line reference measurements assessing novel surface cleaning processes. Grazing-incidence and reference-free XRF employing synchrotron radiation can contribute to the characterization of nanolayered systems without the need for similar reference layers. For the purpose of high-end investigations[1] in the R&D of industrial semiconductor applications, appropriate instrumentation is being further developed by PTB.

---

[1] For information on accessing the PTB instrumentation as well as that of several partner institutes, see www.anna-i3.org, the homepage of the EC-funded integrated infrastructure initiative 'European Integrated Activity of Excellence and Networking for Nano- and Micro-Electronics Analysis' (ANNA).

Acknowledgments

Siltronic AG is gratefully acknowledged for providing Si wafer samples and SiCrystal AG for SiC wafer samples.

References

1. B. Beckhoff, R. Fliegauf, G. Ulm, J. Weser, G. Pepponi, C. Streli, P. Wobrauschek, T. Ehmann, L. Fabry, C. Mantler, S. Pahlke, B. Kanngießer, W. Malzer, *Electrochem. Soc. Proc.* **2003-03,** 120, 2003.
2. C. Streli, P. Wobrauschek, L. Fabry, S. Pahlke, F. Comin, R. Barett, P. Pianetta, K. Lüning, B. Beckhoff, 'Total-Reflection X-Ray Fluorescence (TXRF) Wafer Analysis', *Handbook of Practical X-Ray Fluorescence Analysis* [B. Beckhoff, B. Kanngießer, N. Langhoff, R. Wedell, H. Wolff (Eds.)], Springer, 2006.
3. F. Scholze, B. Beckhoff, M. Kolbe, M. Krumrey, M. Müller, G. Ulm, *Microchim. Acta* **155**, 275, 2006.
4. M. Müller, B. Beckhoff, G. Ulm, B. Kanngießer, *Phys. Rev. A* **74,** 012702, 2006.
5. G. Pepponi, B. Beckhoff, T. Ehmann, G. Ulm, C. Streli, L. Fabry, S. Pahlke, P. Wobrauschek, *Spectrochim. Acta B* **58**, 2245, 2003.
6. S. Török, J. Osan, B. Beckhoff, G. Ulm, *Powder Diffraction J.* **19**, 81, 2004.
7. M. Kolbe, B. Beckhoff, M. Krumrey, G. Ulm, *Spectrochim. Acta B* **60,** 505, 2005.

Structural and Analytical Characterization by Scanning Transmission Electron Microscopy of Silicon-based Nanostructures

A.Armigliato, R.Balboni and A.Parisini

CNR-Istituto IMM, Via P.Gobetti, 101, 40129 Bologna (Italy)

A few recent applications of scanning transmission electron microscopy (STEM) methods to problems of interest for nanoelectronics are reported. They include nanometer-scaled dopant profiles by Z-contrast and strain mapping by convergent beam diffraction.

Introduction

In the present and future CMOS technology, due to the ever shrinking geometries of the electronic devices, the availability of techniques capable of performing a quantitative analysis of the relevant parameters (structural, chemical, mechanical) at a nanoscale is of a paramount importance. The influence of these features on the electrical performances of the nanodevices is a key issue for the nanoelectronics industry (see, e.g. (1))

In this paper it will be reported on the nanoanalysis of two very important physical quantities which need to be controlled in the fabrication processes of nanodevices: the dopant profile in the ultra-shallow Si junctions (USJ) and the lattice strain that generates in the Si electrically active regions of isolation structures. Both these quantities are characterized by methodologies of the scanning transmission electron microscopy (STEM) technique; namely, the dopant profiles are investigated by the so-called Z-contrast annular dark field, (ADF-STEM) method, whereas the mechanical strain will be mapped by the convergent beam electron diffraction (CBED) method. A spatial resolution lower than one nanometer and of a few nanometers can be achieved in the two cases, respectively.

For each of the two methodologies, the paper will first discuss their basic principles. Then, an example of recent applications to the determination of the profile of As, implanted at a low-energy into silicon (Z-contrast), as well as of the two dimensional strain mapping in a shallow-trench isolation (STI) structure (CBED) will be reported.

Dopant Profiles in Silicon by Z-Contrast in HAADF/STEM

Principles of the Method

The Z-contrast ADF-STEM, procedure, applied to the observation of heavily doped implanted Si, was pioneered by S. J. Pennycook and co-workers (2). More recently, this method has been reconsidered by Merli et al. (3,4) and applied to the investigation of ultrashallow junctions in Si using low-energy electrons in a scanning electron microscope, SEM. In those works, the authors arrive at a unified definition of the output signal that in both backscattering electrons BSE and low-energy STEM imaging has a resolution given by the probe size; whereas the signal contrast is found to depend on the interaction volume, in BSE imaging, and on the beam broadening, in STEM. The sensitivity of the method for As and Sb-implanted species was of the order of 1 at. %, while the resolution as defined by the probe size was of the order of 1 nm operating with a SEM equipped with a Schottky emitter at an energy of 20 kV (4). The results reported here (5,6) represent an extension of this approach to the high-energy STEM case, thus coming back to the original Z-contrast application (2), here redefined and optimized to satisfy some of the new needs of the ultrashallow implants characterization. Among these, the dopant localization at a subnanometer scale, at present only possible with high-energy Z-contrast techniques, as well as the determination of the impurity distribution close to the sample surface. As in previous approaches, we are not attempting to resolve dopant atoms but only to detect the increased electron scattering due to the local dopant concentration over the Si matrix albeit keeping this average localization at the subnanometer scale. Key parameters to achieve this selection of a pure Z-contrast signal, i.e., a signal as much as possible independent of the diffraction contrast contributions of the Si matrix, were found to be the electron probe convergence angle (8 mrad), the inner detector angle (62 mrad), and a slight tilt of the sample away from the <011> cross-section zone axis around the <100> surface normal (5,6). Incident probe size dimensions were kept in between 0.2 and 0.3 nm, as demonstrated by lattice images regularly obtained on crystalline regions when tilting the sample back to the on-axis condition. In the following, contrast profiles related to the atomic number difference between dopant and Si matrix atoms have been calculated from line profiles of the ADF-STEM micrographs averaged over regions about 25 nm wide. The contrast is defined in agreement with Elliot et al. (7) as

$$C(t) = \frac{[I(t) - I_{sub}]}{(I_{sub} - I_{ref})}$$

[1]

where $I(t)$ is the averaged intensity at depth t, I_{sub} is the constant Si substrate intensity and I_{ref} is the reference intensity obtained from a beam-blanked image.

Applications to the case of USJ in As implanted Si.

In the case of 5 keV, 2×10^{15} As$^+$/cm^2 implantations in Si, the occurrence of an As diffusion towards the surface and a consequent dopant pileup in the surface region has been suggested by secondary ion mass spectroscopy, SIMS, measurements after annealings at 800 °C. However, it is known that SIMS measurements could be affected by some artifacts in proximity of the surface or of the interface with the surface oxide.
To verify the presence of a surface dopant accumulation, ADF-STEM cross-sectional observations have been carried out as preliminarily reported in (5). In Fig. 1a and b, it is shown the typical aspect of the ADF-STEM micrographs taken in on-axis and off-axis sample orientations, respectively. The comparison between these micrographs, taken on the same region of an as-implanted sample, also shows how effectively the diffraction and channeling contrast can be removed by the ADF-STEM micrographs with a slight tilt of the sample. In fact, in the case of the tilted sample, Fig. 1b, a broad intensity maximum is uniquely observed close to the sample surface and even the interface between the amorphized surface region and the Si crystal, Fig. 1a, is no longer visible. In Fig. 2a and b, ADF-STEM micrographs obtained on the implanted samples before and after an annealing at 800 °C for 3 min, respectively, are reported. Three regions can be distinguished based on the visible intensity variations. Starting from the sample surface at the top of the micrographs, a 2 nm wide region corresponding to the Si oxide layer is observed, followed by a brighter doped region whose intensity gradually vanishes, at a depth of about 15 nm, into the darker Si matrix region.

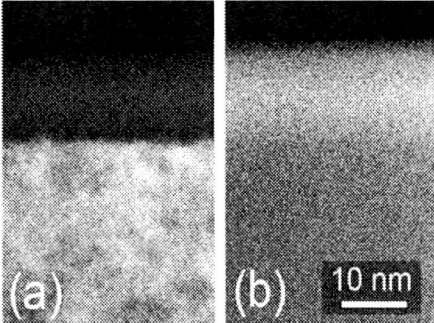

Fig. 1. Z-contrast ADF-STEM micrographs obtained on the same region of a 5 keV $2x10^{15}$ As^+/cm^2 implanted Si sample in on-axis (a) and off-axis (b) sample orientations.

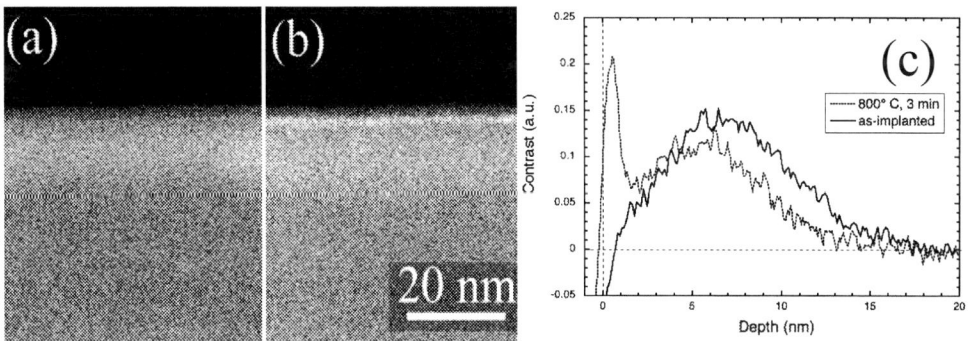

Fig. 2. ADF-STEM micrographs (a, b) and corresponding contrast profiles (c) obtained on 5 keV $2x10^{15}$ As^+/cm^2 implanted Si samples before (a) and after (b) annealing at 800 °C for 3 min.

The contrast profiles reported in Fig. 2c evidence that after the annealing at 800 ° C for 3 min, a pronounced contrast peak close to the sample surface is present. These profiles have been obtained from regions of equal thickness, a relative sample thickness of about 0.7 t/λ (λ is the inelastic electron mean free path, that for pure Si and in our experimental situation corresponds to about 100 nm) being measured in both the cases by parallel electron energy loss spectroscopy, PEELS. Although a detailed investigation of the dependence of these results on the TEM sample thickness has not yet been achieved, we have observed that the effects of the beam broadening on the visibility of the dopant profile become evident for relative specimen thickness $t/\lambda \geq 1.5$. Possible artifacts of the TEM sample preparation have also been excluded by PEELS analyses showing that the surface peak, systematically observed in these experimental conditions, did not correspond to a thicker sample region. The surface contrast peak is thus interpreted as an actual dopant accumulation, confirming the presence of an uphill As diffusion phenomenon. The agreement of both the profiles with the corresponding SIMS profiles is remarkable (5,6) and seems to indicate that, in this case, the sensitivity of the technique is of the order of, or slightly better than, 1 at. %. Finally, it is worth noting that in Fig. 1c, the areas underneath the as-implanted and annealed sample profiles differ by no more than 10%, indicating a constant As dose, at least within the experimental errors. This means that even in the case of ultrashallow junctions, a standardless and quantitative analysis of impurity concentration should be possible, as originally suggested for shallow implants (2).

Strain Mapping by Convergent Beam Electron Diffraction

Principles of the Method

The convergent beam electron diffraction (CBED) technique of the transmission electron microscopy is presently the only method capable of yielding quantitative strain information with a resolution at the nanometer scale; it is a point-to-point technique, which allows the strain tensor to be obtained at each nanoregion of the sample probed by the electron beam from the analysis of the corresponding diffraction pattern. Basically, the CBED method is based on the strain induced shift of High Order Laue Zones (HOLZ) deficiency lines (8), which occur in the central disk of a convergent-beam pattern, taken in a zone axis where the HOLZ lines are free from dynamical interactions . Due to their high-angle scattering origin, the position of these lines is very sensitive, among other parameters, to small variations in acceleration voltage and lattice parameters (strain). An example of a strain induced HOLZ line shift is given in Fig. 3, which refers to the case of a TEM cross section of a silicon wafer covered with a Si-10 at % Ge film. The HOLZ line pattern taken in the undeformed silicon substrate (left) is clearly different from that taken in the region of the Si-Ge alloy (right). The coherent growth of the film on the substrate results in a tetragonal distortion $\varepsilon_T=6x10^{-3}$. Strains of the order of $2x10^{-4}$ can give rise to a detectable HOLZ lines shift.

In order to quantify the strain, it is first necessary to assess the effective acceleration voltage by matching the pattern taken on the unstrained part of the sample with a kinematically simulated one (9). Then the unknown lattice parameters are determined by fitting the experimental pattern taken on the strained layer with a simulated one, assuming the lattice constants as fitting parameters.

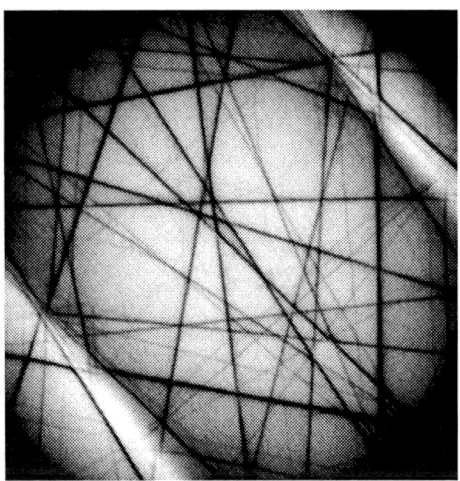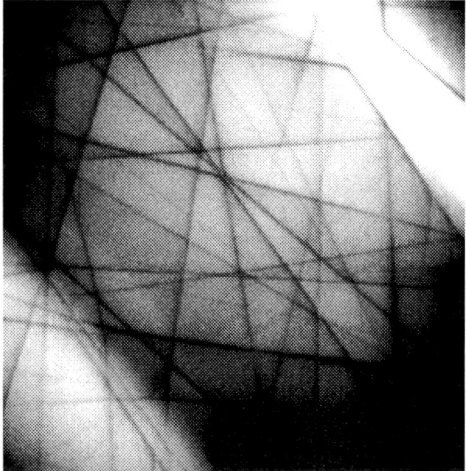

Figure 3. Strain analysis in silicon. Comparison of CBED patterns taken in undeformed silicon (left) and in a highly deformed area of a sample (right). The method is based on the measurement of the shift in the position of the diffraction lines.

This involves first the determination of the position of the HOLZ lines in the experimental patterns, then the extraction of strain tensor from each pattern by comparing it with simulations. The first task is presently accomplished using a dedicated routine (ASAC) in the software iTEM® (by Olympus-SIS (10), which has been developed in the framework of a European project (11), on the basis of the HOLZFIT programme set up in our Institute. The software detects the HOLZ lines by using a formalism similar to the Hough transforms, i.e. all the lines detected in the pattern are represented in a transformed space by their distance from the reference origin and the angle they make with a reference axis. Each pattern is then unambiguously parameterised by computing a set of relevant distances between a number of intersections of the detected HOLZ lines (Fig. 4).

The extraction of the strain tensor is made by comparison of the experimental distance set with the corresponding one obtained by a quasi-kinematical simulation of CBED patterns; the procedure is fully automated and the strained structure which best matches the experimental pattern is extracted using a χ^2 minimisation criterion (12). The effective voltage is calculated by collecting a CBED pattern in the undeformed substrate and simulating it using the voltage as the fitting parameter; then the same voltage is used for the strain tensor calculations. By doing this, all the six lattice parameters are then extracted from a single CBED pattern, as the output of the minimisation routine. The strain tensor component are finally calculated using Equations [2]:

$$\varepsilon_{XX} = \frac{a_S^X - a_0}{a_0} \qquad \varepsilon_{XY} = \frac{1}{2}\left(\frac{\pi}{2} - \gamma_S\right)$$

$$\varepsilon_{YY} = \frac{a_S^Y - a_0}{a_0} \qquad \varepsilon_{XZ} = \frac{1}{2}\left(\frac{\pi}{2} - \beta_S\right) \qquad [2]$$

$$\varepsilon_{ZZ} = \frac{a_S^Z - a_0}{a_0} \qquad \varepsilon_{YZ} = \frac{1}{2}\left(\frac{\pi}{2} - \alpha_S\right)$$

where X,Y,Z are the crystallographic axes, S means strained and 0 undeformed silicon. From these components, the tensor trace Tr(ε) can be simply obtained, as Tr(ε)=ε_{XX}+ε_{YY}+ε_{ZZ}.

In the practical work, the number of unknown parameters must be reduced from 6 (*a*, *b*, *c*, *α*, *β*, *γ*) to 3, to obtain a unique solution. As the cross-section orientation is along the [1 1 0] direction, the relations *a=b* and $\Delta\alpha = -\Delta\beta$ hold. In addition, for the specimen thickness (200-300 nm) used in our CBED experiments, the so called 'planar strain' approximation (i.e. a negligible relaxation in the direction perpendicular to the cross section plane) can be assumed, so $\Delta a/a = \Delta\gamma/2$. A detailed explanation of these assumptions has been reported elsewhere (13).

The spatial resolution of the CBED technique is in principle given by the spot size (about 1 nm). This holds along the [0 0 1] direction of the cross-sectioned TEM sample, whereas, due to the sample tilting used in the CBED experiments, the spatial resolution in the [1 -1 0] direction of the TEM cross section perpendicular to the [0 0 1] tilt axis is worsened by the projection effect, which increases with the specimen thickness. For instance, in the case of Fig. 3, as the <230> zone axis is 11.3° off the <110> normal to the plane of the cross section, the spatial resolution in the [1 -1 0] direction is about 10%

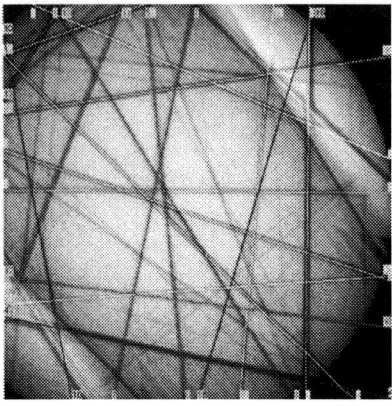

Figure 4. Undeformed silicon CBED pattern (<230>, 200 kV as in Fig. 2); the HOLZ lines skeleton, detected by the ASAC/iTEM® software, is shown superimposed to the experimental lines.

of the local sample thickness, i.e. 20 nm in a 200 nm-thick region of the device. For this reason, it is more convenient to reduce the angle of tilt, and the <340> zone axis (ca. 8° of tilt) has been recently introduced, thus improving the spatial resolution along the [1 -1 0] direction by about 35% with respect to the <230> axis.

Application of STEM/CBED to Strain Maps

The above described method has been so far applied to mapping strain in shallow trench isolation technology of the recent CMOS technology nodes (150-90 nm). The example given here refers to an STI with an active region about 150 nm wide (Fig. 5) and has been chosen to show more clearly the array of analyzed points and of the corresponding values of the trace.

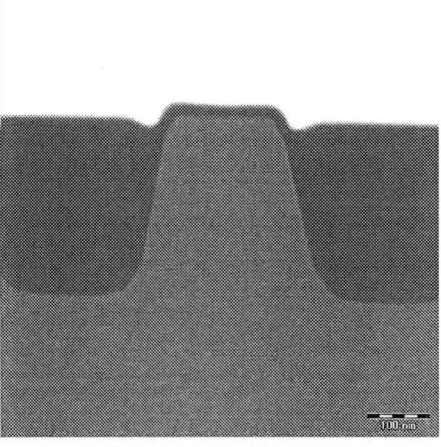

Figure 5. HAADF/STEM image in <110> Si zone axis of an STI structure.

A number of CBED patterns has been taken in points selected by digitally rastering the probe in the 2D region of interest of the cross section; in addition, a CBED pattern in an undeformed area of the substrate is acquired to determine the effective acceleration voltage (9). A database is obtained, consisting of a TEM image with superimposed the matrix of the points selected for CBED pattern acquisition, plus the CBED patterns obtained at each point. From the analysis of each pattern, the local strain tensor is obtained following the above mentioned procedure. An example of the map of the tensor trace is given in Fig. 6. There are presently two ways of visually displaying the strain information, by associating to each investigated point the corresponding value of the component of the strain tensor (the trace in this case). In Fig. 6a the numerical value of the trace (in 1E-4 units) is shown, whereas the geometrical shape of the active region of the STI is superimposed to the two-dimensional map just to guide the eye. The data in Fig. 6b are deduced from exactly the same output file of ASAC as in Fig. 6a, but now the dots are given in colours, according to a selected palette (here blue and black indicate large and small compressive strains, respectively); to the map is superimposed the corresponding ADF-STEM micrograph. In the same way all the components of the strain tensor can be plotted as two-dimensional maps.

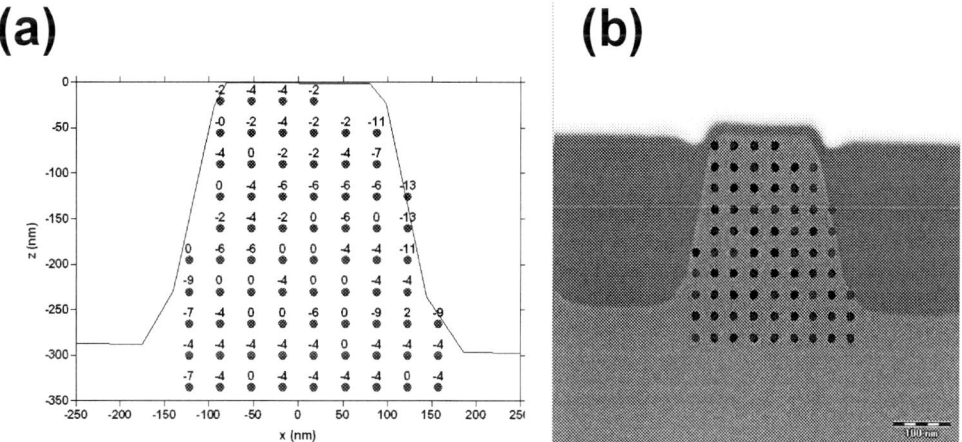

Figure 6. Maps of the tensor trace of the STI structure in Fig. 5. Measurements were performed in <340> orientation. In order to compare maps of active regions in differently processed STIs, each point can be associated with the corresponding numerical strain value (as in (a), 1E-4 units) or with different colours (as in (b), where blue and black correspond to large and small compressive strains, respectively). In (a) the profile of the shallow trenches is drawn to guide the eye. Note that by the ASAC software all the components of the strain tensor can be mapped. Adapted from (14).

Conclusions

The two STEM-based techniques described in this work (Z-contrast ADF and CBED) have proved to be very powerful in the quantitative analysis of dopant profiles in ultra-shallow junctions in silicon and of lattice strain mapping in STI structures, respectively. Their high spatial resolution and sensitivity make them very useful in the intercomparison of results obtained by different techniques, which is a key issue of the European project ANNA. In particular, it is planned to compare the dopant profiles with those obtained by SIMS (Secondary Ion Mass Spectrometry) and MEIS (Medium Energy Ion Scattering) advanced methodologies; likewise, the strain values obtained by CBED will be compared with the two dimensional process simulations currently employed in CMOS technology and the impact of strain on the electrical performances of the devices will be investigated.

Acknowledgments

This work is partially supported by the project ANNA (European Integrated Activity of Excellence and Networking for Nano- and Micro-Electronics Analysis) of the I3 initiative of the EU 6th Framework Programme.

References

1. P. Smeys, P.B. Griffin, Z.U. Rek, I. De Wolf and K.C. Saraswat, *IEEE Trans. Electron Dev.*, **46**, 1245 (1999)
2. S.J. Pennycook, S. D. Berger, and R. J. Culbertson, *J. Microsc.*, **144**, 229 (1986).
3. P.G. Merli, V. Morandi, and F. Corticelli, *Appl. Phys. Lett.*, **81**, 4535 (2002).
4. P.G. Merli, V. Morandi, G. Savini, M. Ferroni, and G. Sberveglieri, *Appl. Phys. Lett.*, **86**, 101916 (2005).
5. M. Ferri, S. Solmi, A. Parisini, M. Bersani, D. Giubertoni and M. Barozzi, *J. Appl. Phys.*, **99**, 113508 (2006).
6. A. Parisini, D. Giubertoni, M. Bersani, M. Ferri, V. Morandi, P. G. Merli, Proceedings of the 8th Multinational Congress on Microscopy, Prague, 18-21 june 2007, p. 43.
7. S.L. Elliott, R.F. Broom, and C.J. Humphreys, *J. Appl. Phys.*, 91, 9116 2002.
8. P.M.Jones, G.M.Rackham and J.W.Steeds, *Proc. R. Soc. Lond.*, **A 354**, 197 (1977).
9. Y.P. Lin, D.M. Bird and R. Vincent, *Ultramicroscopy*, **27**, 233 (1989).
10. ASAC: http://www.soft-imaging.com/en/164_190.htm
11. STREAM Contract No. IST-1999-10341 (http://stream.bo.cnr.it)
12. R.Balboni, S.Frabboni and A.Armigliato, *Phil. Mag.*, **A77**, 67 (1998)
13. Deliverable D2, STREAM project, p.25 (2000) (http://stream.bo.cnr.it)
14. A.Armigliato, R.Balboni and A.Frabboni, *Appl. Phys. Lett.*, **86**, 63508 (2005).

Thin Film Analysis and Model Interface Characterization Studies of Relevance to Microelectronics

I. Dontas[a], V. Papaefthimiou[a], S. Kennou[a], and S. Ladas[a]

[a] Surface Science Laboratory (SSL), Department of Chemical Engineering, University of Patras & FORTH/ICE-HT, GR-26504, Rion, Patras, Greece

In order to demonstrate the application of surface sensitive techniques in microelectronics-related model interface studies, two such examples of recent work at the SSL are briefly presented. The first involves the controlled growth of the interface between a conjugated organic semiconducting film (Ooct-OPV5, an oligomer of PPV) and p-doped Si(111), which is relevant to hybrid organic/ inorganic microelectronic devices. The second example concerns the comparative study of *in situ* grown metal/6H-SiC (0001) interfaces using a series of metals with different work functions, which is relevant to Schottky contact formation in devices. Using a combination of spectroscopic and other techniques in each case, information concerning film composition and electronic structure, interfacial chemistry, band bending and interface dipole formation are obtained, leading to band line-up diagrams, from which band offsets or carrier injection barriers can be estimated.

Introduction

The use of a wide range of surface sensitive techniques for the analysis/characterization of ultra-thin films and interfaces is well established and documented (1). The main emphasis at SSL (2) in the frame of the ANNA project (3) is to make versatile combinations of a number of such techniques, all clustered around the same ultra-high-vacuum (UHV) chamber, in order to study systems of interest to microelectronic device technology. These techniques include : X-ray photoelectron spectroscopy (XPS) and Auger electron spectroscopy (AES) for elemental analysis with chemical state identification over a depth range of a few nm, low-energy ion scattering spectroscopy (LEISS) for elemental analysis and structural characterization of the outermost atomic surface layers, ultra-violet photoelectron spectroscopy (UPS) and electron energy loss spectroscopy (EELS) for surface bonding and electronic structure determination, including work function (WF) measurements by UPS, low-energy electron diffraction (LEED) for surface structure / epitaxial characterization in single-crystalline films, contact-potential difference (CPD) measurements using a vibrating Au reference electrode for determining WF changes and mass spectroscopic techniques, like temperature programmed desorption/reaction (TPD/TPR), to study gas/solid interactions and surface chemical reactivity. All the above techniques are basically non-destructive, however, XPS / AES can be used in combination with controlled removal of successive surface layers by inert ion sputtering in order to yield analytical depth-profiling over a practical depth range of 100nm.

There are two principal ways of applying these techniques in order to obtain useful information for semiconductor materials and device characterization. The first approach involves direct surface analysis and/or depth profiling on specimens provided by their manufacturers. These specimens are analysed as-received or after a mild sputtering treatment to remove superficial contamination layers arising from their exposure to the atmosphere, unless of course the surface contamination is in itself the object of analysis. The second approach is to create *in situ*, under model, controlled conditions the desired systems for analysis and characterization. Recent studies at the SSL involve the gradual growth of selected interfaces between semiconductor, metal or insulator materials, usually by physical vapor deposition of one material on top of a well characterized substrate inside the same UHV analysis chamber. Two examples of such model interface studies are going to be briefly described. The first involves the controlled growth of the interface between a conjugated organic semiconducting film (Ooct-OPV5, an oligomer of PPV) and an inorganic substrate - p-doped Si(111), which is relevant to hybrid organic / inorganic microelectronic devices. The second example concerns the growth of metal/6H-SiC (0001) interfaces using metals with considerably different work functions, which is relevant to Schottky contact formation in devices. Using a combination of spectroscopic or other techniques in each case, information concerning film composition and electronic structure, interfacial chemistry, band bending and interface dipole formation are obtained, as well as energy diagrams, from which band offsets or carrier injection barriers can be estimated.

The Si(111)/ Ooct-OPV5 interface

The use of organic/silicon hybrid structures constitutes a promising extension of traditional microelectronic technology into new types of devices, therefore it has received recently growing attention (4). The study of the interfaces between organic layers and silicon electrodes is indispensable, as the interfacial physical and chemical phenomena determine the efficiency of device operation. Photoelectron spectroscopic techniques, like XPS and UPS , are well suited for that purpose. The results reported here concentrate on the system Ooct-OPV5 /Si(111), whereby the organic oligomer is a model compound for the widely used poly-(*p*-phenylenevinylene) PPV (5) and the substrate is *p*-doped Si (resistivity 11-22 Ohm cm).

Experimental

The spectroscopic measurements were performed in a conventional ultra high vacuum (UHV) chamber (base pressure 6×10^{-10} mbar), which has been described elsewhere (6). The substrate was chemically cleaned *ex situ* (7) and then subjected to argon ion sputtering and annealing up to 1150 K inside the UHV chamber, so that residual C and O contamination was reduced to a few tenths of a monolayer. Thin Ooct-OPV5 films up to 10 nm were produced inside the UHV chamber by step-wise sublimation from a home-made vapor deposition source (6). The growth of the interface was followed by XPS (AlKα at 1486.6 eV) and UPS (HeI at 21.2 eV), whereby the analyser energy scale was calibrated against a binding energy value of 84.00 ± 0.05 eV for the Au $4f_{7/2}$ peak of a sputter-cleaned Au foil and the approximate instrumental resolution was 1.5 eV for XPS and 0.16 eV for UPS. The average thickness of the oligomer film was obtained from the attenuation of the substrate peaks upon deposition assuming a layer-type growth.

Results and Discussion

Figure 1 shows the variation of the C1s XPS peak arising from the growing oligomer film and the variation of the Si 2p peak from the substrate upon deposition. In all cases the peaks have been analysed into a number of components. The Si 2p peak, which contains a small higher binding energy (BE) contribution from Si atoms associated with residual surface oxygen, is attenuated upon film growth but its position and shape remain unchanged. The measured Si 2p BE of 99.7 eV is about 0.65eV higher than that expected under flat band conditions for the p-doping level of the Si substrate (7.5×10^{15} cm^{-3}), which suggests that at the clean substrate surface there exists a strong downward band bending which pins the Fermi level 0.85 eV above the VB maximum close to the conduction band edge. The growing C1s peaks are normalized to the same height so that the details of their shape and position are more clear and they are analysed into three components, reflecting the contribution of carbon atoms from various functional groups of the Ooct-OPV5 molecule, main chain, alkyl side chains and ether groups in ascending BE order (8). The relative intensities of these peaks are representative of the oligomer stoichiometry even for submonolayer average film thickness (less than 1 nm) indicating negligible chemical interaction and an abrupt interface. After the first 2 monolayers of deposited oligomer, the C1s BE begins to shift towards higher values up

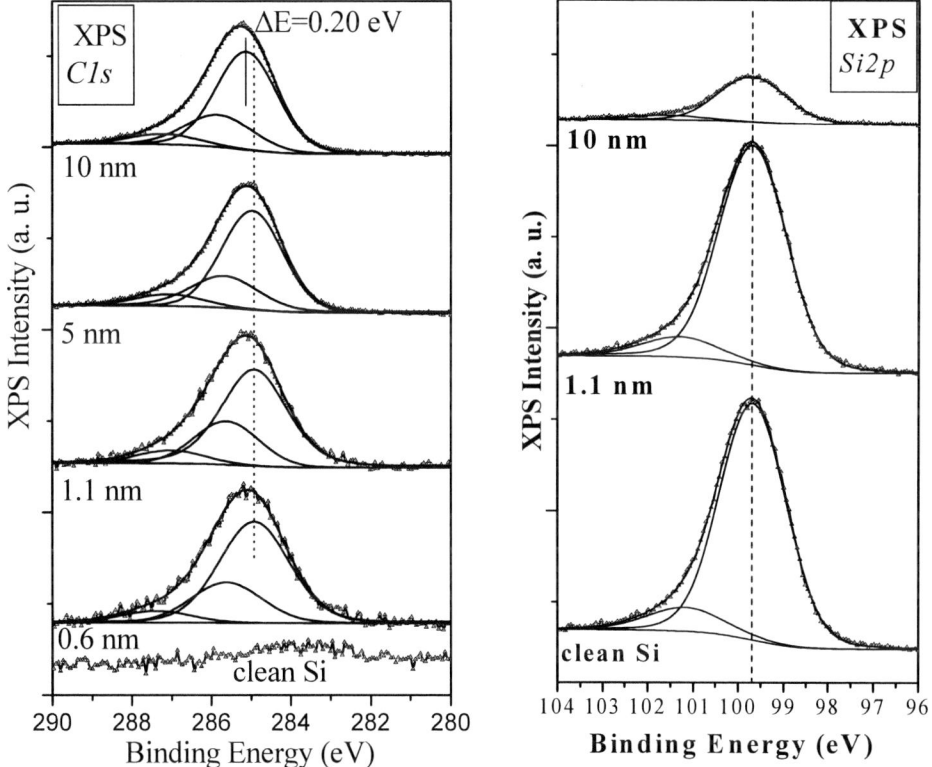

Figure 1. (Left): Evolution of the normalized C1s peak upon oligomer film growth. (Right): Evolution of the Si2p substrate peak upon oligomer film growth.

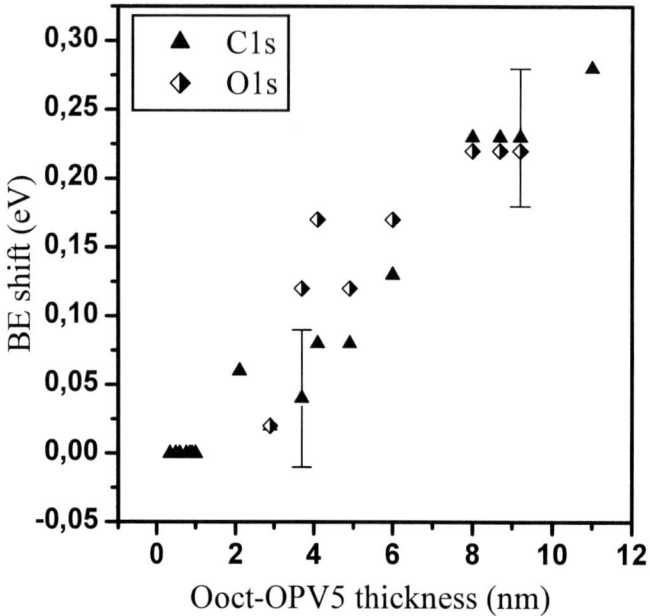

Figure 2. Variation of the BE for oligomer related peaks as a function of deposited thickness.

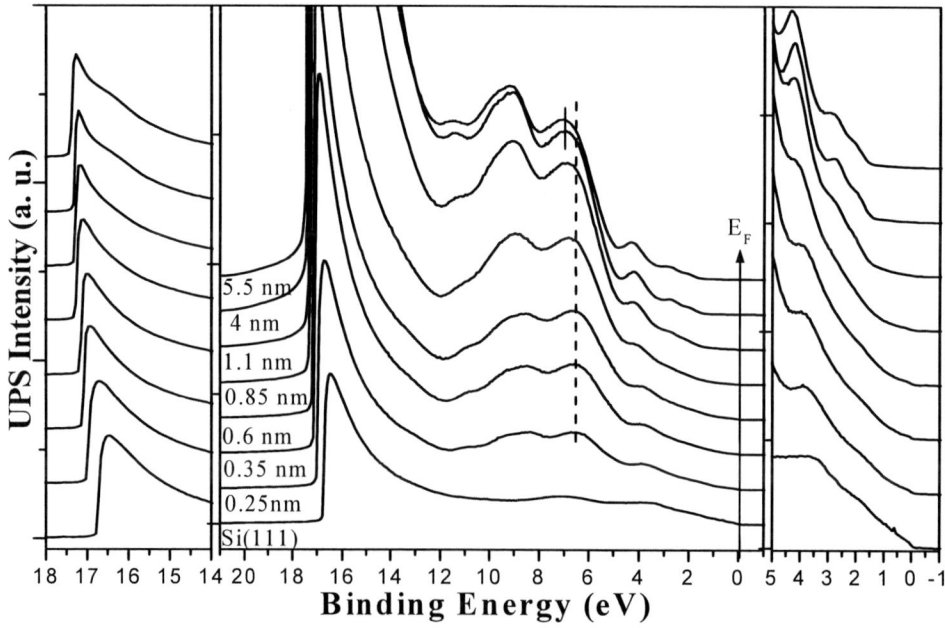

Figure 3. The development of UP spectra upon oligomer deposition on the Si substrate.

to a total shift of about 0.20 eV around a film thickness of 8 nm. This shift is also exhibited by the O 1s peak from the oligomer and is shown in fig. 2 as a function of average film thickness. Note that the further upward shift beyond a thickness of 10 nm in fig.2 is due to the onset of electrostatic charging in the photoemission experiments due to the limited conductivity of the film. The binding energy shift is not related to final state screening effects, since it appears only after at least two monolayers of oligomer deposition. It is attributed to band bending exclusively in the growing semiconducting film of the Ooct-OPV5, since the substrate does not exhibit any changes in its original band bending state, as evidenced by the constancy of the Si 2p BE (fig.1). The 0.20 eV band bending value over a depletion zone of the order of 8 nm for Ooct-OPV5, with a dielectric constant $\varepsilon_r \sim 3$, suggests a charge carrier concentration of the order of 10^{18} cm^{-3}, in fair agreement with published results (9).

Figure 3 shows a sequence of UP spectra upon growth of Ooct-OPV5 on the Si substrate. The oligomer valence band features appear upon the first few deposition steps and eventually dominate the spectrum which resembles strongly that of PPV (10). The highest occupied molecular orbital (HOMO) cut-off position of the bulk Ooct-OPV5 film was measured 1.45±0.05 eV from the analyser E_F (right part of fig.3) in agreement with a previously measured value (6). From the variation of the low energy cut-off edge of the UP spectra (left part of fig.3) one can follow the work function changes upon oligomer film growth. Also by subtracting the UP spectrum width from the HeI excitation energy (21.2 eV) one can obtain the absolute value of the work function, which for the clean substrate is 4.50 eV. Within the first 1 or 2 monolayers of deposited oligomer the work function has reached the value for the bulk Ooct-OPV5 at 4.00 eV (6), indicating a layer-type growth.

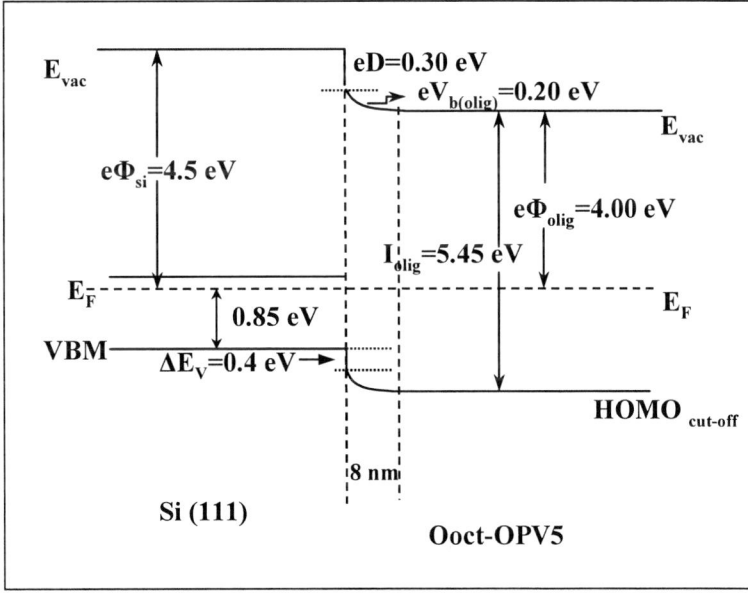

Figure 4. Band line-up diagram at the interface between p-doped Si(111) and a thick Ooct-OPV5 film. Note that at the surface of the Si substrate the Fermi level has been pinned close to the conduction band.

By combining the information obtained from the spectroscopic results, it is possible to construct the band line-up diagram shown in Figure 4. The value of the interface dipole eD=0.30 eV is obtained by subtracting the band bending in the film from the total decrease of the work function. The ionization energy of the oligomer, which is obtained as the sum of the film work function and the HOMO cut-off, is 5.45 eV in agreement with previous results (6). The most interesting parameter obtained from this diagram is the valence band offset at the interface, which acts as a hole injection barrier under proper biasing. This value is 0.40 eV for the above system and would be a useful input from this model study towards the design or the description of the corresponding device operation.

Comparative study of metal / 6H-SiC (0001) contacts

Metal/SiC interfaces are important for high temperature and high power microelectronic devices because of the high breakdown field and the wide band gap of SiC, therefore the electrical behavior and the thermal stability of such contacts have attracted particular attention (11-13). Metal/SiC contact formation has been studied for a large number of metals in recent years at room temperature and in some cases the effect of higher temperature annealing on the stability of the interface has been also investigated (14-18). The Schottky barrier height (SBH) of such contacts has been measured in model studies, whereby the metal films were *in situ* grown in UHV and the SBH was determined by XPS following the method developed by Waldrop and co-workers (14), as well as by using conventional electrical measurements (I-V, C-V). The interfaces of 6H-SiC (0001) surfaces (Si- or C-terminated) with Re, Er and Cu and their thermal stability have been the subject of model interface studies at the SSL in the recent years (19-21), whereas that with Cr is currently under investigation. In this work, some recent results on the Cr/6H-SiC(0001) interface formation, obtained by X-ray photoelectron spectroscopy (XPS), low energy electron diffraction (LEED) and work function (WF) measurements, will first be presented as a demonstration of the experimental approach followed in all model studies. Then, a comparative discussion of the results for the Re, Er, Cu and Cr metal contacts with 6H-SiC(0001) will be made, concentrating in the magnitude of the SBH height and the interface dipole.

Experimental

All measurements were performed in a second ultra high vacuum chamber (base pressure 9×10^{-10} mbar) equipped with a Leybold EA-11 hemispherical electron energy analyser and a twin-anode X-ray gun for XPS. Photoelectrons were excited using the un-monochromatized AlKα line at 1486.6 eV or MgKα line at 1253.6 eV. A Kelvin probe with a gold-plated vibrating reference electrode (piezoelectrically driven) was used to monitor WF changes. Single crystal 6H-SiC wafers, Si-face, from Cree Research Inc. were used as substrates. The wafers were n-type, nitrogen-doped (3×10^{17} - 3×10^{18} cm^{-3} net doping density) and oriented 3-4° off-axis with respect to the ideal (0001) plane towards the [11-20] direction. Before the samples were mounted onto the ultra high vacuum system, they were subjected to standard chemical cleaning to remove the native oxide. In UHV, the specimens were only heated up to 1000 K, since both sputtering and high temperature annealing disturb the surface structure and stoichiometry of the substrate and as a result the surface conditions would not correspond to those actually prevailing during preparation of such contacts in device manufacture. Chromium was evaporated in UHV from a home-built evaporation source. After each deposition step, XPS and WF measurements were taken. After the final deposition of chromium at room temperature, the SBH was determined from the XPS data. The chromium coverage, in equivalent monolayers, was estimated by calibrating the source doses from the exponential decay of the Si2p core level peak intensity, assuming a 2-D growth at the early deposition stages. One monolayer (ML) corresponds to an average film thickness of 0.26 nm, or a Cr surface atom density of $2x10^{15}$ atoms cm^{-2}.

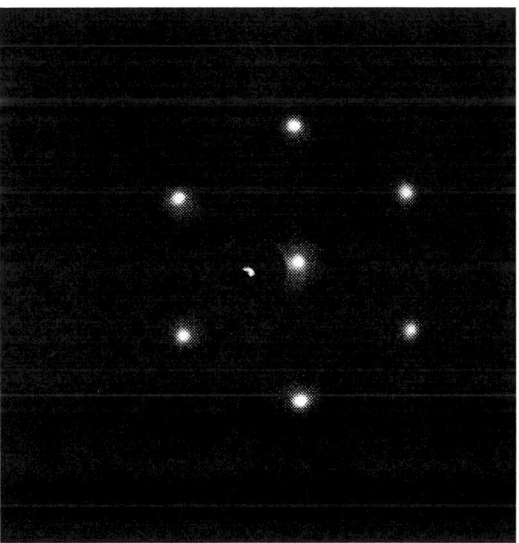

Figure 5. A LEED image of the annealed 6H-SiC(0001) surface prior to deposition (E=190 eV). The [11-20] direction points horizontally to the right .

Results and Discussion

The annealed substrates exhibited a good 1x1 hexagonal LEED pattern, as shown in fig.5. A small O1s peak was observed in the XP spectrum of the annealed sample surface indicating a residual oxygen coverage of about half a monolayer. The work function of the surface prior to metal deposition was taken as 4.4 eV (19-21). Figure 6 shows the evolution of Si 2p and Cr 2p peaks upon metal deposition. The substrate peak is practically unchanged in shape and position, whereas the chromium peak shifts gradually from higher BE towards the expected value for metallic Cr (574.3 eV). The work function (not shown) initially decreases with increasing metal coverage up to ~ 1 monolayer indicating an interfacial interaction limited to the first atomic layer (most likely via the residual surface oxygen) and then increases up to the value corresponding to metallic Cr.

The SBH for the Cr/SiC interface was calculated according to ref.14. In brief, the SBH is equal to E_g - E_{fi}, where E_g = 3.03 eV is the 6H-SiC band gap (14) and E_{fi} is the interface Fermi level measured by XPS with respect to the top of the silicon carbide valence band as $E_{C\,1s}$ − $(E_{C\,1s} − E_V)_o$, where the quantity in parentheses is the binding energy difference C1s to valence-band maximum for the clean SiC, as shown in fig.7.

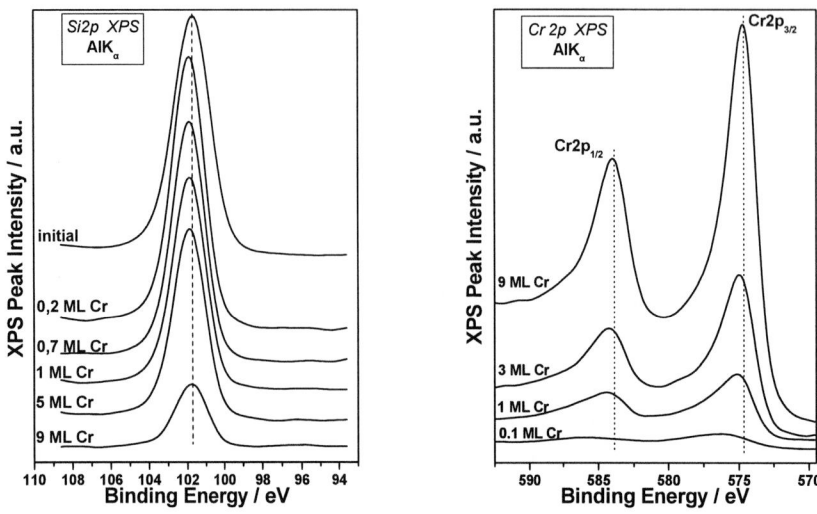

Figure 6. The Si 2p and Cr 2p XPS peak evolution upon metal deposition.

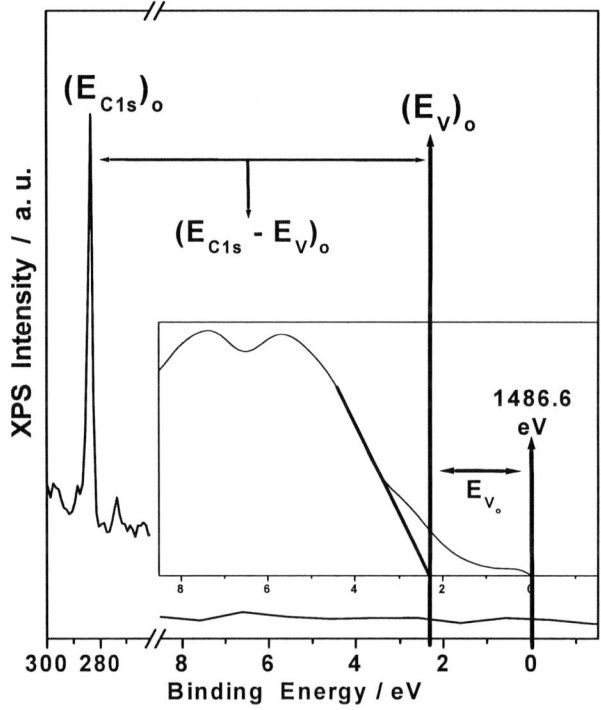

Figure 7. Calculation of the difference between the C1s position and the top of the substrate valence band prior to deposition.

The latter was measured 281.3±0.1 eV in agreement with previously measured values (19-21). From the BE value of C1s after the final Cr deposition (283.4 eV), a SBH of 0.9±0.1 eV is obtained. The only *ex-situ* measured value for the SBH that could be found in the literature for the Cr/6H-SiC contact was 1.15 eV (22).

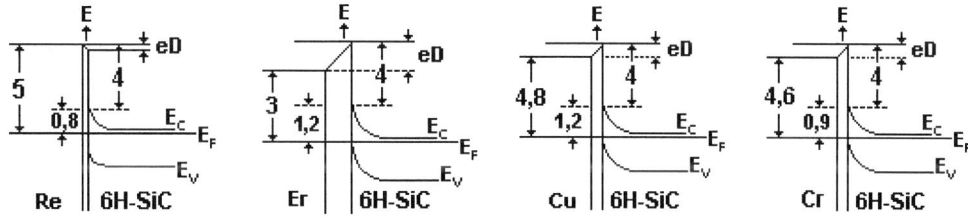

Figure 8. Schematic energy diagrams of the Re, Er, Cu and Cr contacts on 6H-SiC (0001)

The schematic energy diagram for the Cr on n-type 6H-SiC contact is shown in fig.8, along with the corresponding diagrams for the three metals (Re, Er, Cu) previously studied at the SSL on the same substrate (19-21). In order to construct these diagrams , a value of 4 eV for the electron affinity, χ, of the substrate was adopted (14) and literature values for the work function of the metals, $e\Phi_m$, in their polycrystalline form were taken (14, 19-22). The interface dipole , eD , could then be obtained in each case according to the following equation:

$$eD = SBH + \chi - e\Phi_m \qquad [1]$$

The results are summarized in Table I.

TABLE I. Schottky Barrier Height and Interface Dipole for various metal contacts with n-type 6H-SiC(0001).

Metal (Work function / eV)		SBH / eV	Interface Dipole / eV
Re	(5.0)	0.8	~0
Er	(3.0)	1.2	2.2
Cu	(4.8)	1.2	0.4
Cr	(4.6)	0.9	0.3

The results clearly indicate that the simple Schottky-Mott rule, whereby the SBH is equal to the difference between the metal work function and the electron affinity of the semiconductor is not generally applicable, as evidenced by the presence of a non-zero value for the interface dipole in most cases. The interface dipole is usually the result of interface defect states arising from local metal-semiconductor interactions , whereby the presence of any impurities like residual oxygen would play a role. What should be emphasized here is that model experiments, as these described above, can yield reliable values for the Schottky barrier height, which may be compared with the results of electrical measurements on practical contacts and contribute to the understanding of their behaviour.

Acknowledgments

Part of the work presented for the metal/6H-SiC contacts was funded by the European Social Fund (ESF), Operational Program for Educational and Vocational Training II (EPEAEK II) and particularly the Program PYTHAGORAS. Financial support for presenting this work at ALTECH 2007 by the European Commission - Research Infrastructure Action, under the FP6-Program "Structuring the European Research Area", through the Integrated Infrastructure Initiative "European Integrated Activity of Excellence and Networking for Nano and Micro- Electronics Analysis", contract n. 026134(RII3)ANNA is also acknowledged.

References

1. M. P. Seah, in *Practical Surface Analysis,* D. Briggs and M. P. Seah Editors, Volume I, Wiley , New York (1990).
2. http://athena4.chemeng.upatras.gr
3. http://www.anna-i3.org
4. S. F. Bent, *Science* , **500**, 879 (2002).
5. R. E. Gill, A. Meetsma and G. Hadziioannou, *Adv. Mater.(Weinheim Ger.)* , **8**, 212 (1996).
6. A. Siokou, V. Papaefthimiou and S. Kennou, *Surf. Sci.*, **482**, 1186 (2001).
7. A. Ishizaka, K. Nakagawa and Y. Shiraki, *Proceedings of the Second International Symposium MBE-CST 2*, 183 (1985).
8. D. Briggs, *Surface Analysis of Polymers by XPS and Static SIMS* , Cambridge University Press, Cambridge UK (1998).
9. S. C. Jain, W. Geens, A. Mehra, V. Kumar, T. Aernouts, J. Poortmans, R. Mertens and M. Villander, *J. Appl. Phys.*, **89**, 3804 (2001).
10. M. Logdlund and W. R. Salaneck, *Surf. Sci. Spectra*, **3**, 384 (1997).
11. J. W. Palmour, J. A. Edmond, H. S. Kong and C. H. Carter, *Physica B*, **185**, 461 (1993).
12. H. Morkoc, S. Strite, G. B. Gao, M. E. Lin and M. Burns, *J. Appl. Phys.*, **76(3)**, 1363 (1994).
13. P. R. Chalker, *Thin Solid Films*, **343-344**, 616 (1999).
14. J. R. Waldrop, R. W. Grant, Y. C. Wang and R. F. Davis, *J. Appl. Phys.*, **72(10)**, 4757 (1992).
15. L. Li and I. S. T. Tsong, *Surf. Sci.* , **364**, 54 (1996).
16. I. Shalish and Y. Shapira, *IEEE Electron Device Lett.*, **21(12)**, 581 (2000).
17. T. Suezaki, K. Kawahito, T. Hatayama and T. Fuyuki, *Jpn. J. Appl. Phys Lett.*, **240(1A/B)**, L43 (2001).
18. F. La Via, F. Roccaforte, A. Makhtari and L. Calcagno, *Microelectron. Eng.*, **60**, 269 (2002).
19. S. Kennou, A. Siokou, I. Dontas and S. Ladas, *Diamond Rel. Mater.*, **6**, 1424 (1997).
20. I. Dontas and S. Kennou, *Diamond Rel. Mater.*, **10**, 13 (2001).
21. I. Dontas, S. Ladas and S. Kennou, *Diamond Rel. Mater.*, **12**, 1209 (2003).
22. S.Yu. Davydov, A. A. Lebedev, O.V. Posrednik and Yu. M. Tairov, *Semiconductors*, **35(12)**, 1375 (2001) (Translated from Russian).

Surface Microdefects Control during Chemical Mechanical Polishing of Silicon Wafers: an Example of in line Manufacturing Process Control

G.Borionetti, A.Corradi, N.Mainardi, A.M.Rinaldi, K.Takami,

[a] MEMC Electronic Materials SpA Viale Gherzi,, 31 28100 NOVARA Italy

For the process control / improvement in Chemical Mechanical Polishing in silicon wafer production line, the automatic laser inspection tool (SP1) and on-time manufacturing database system are utilized for monitoring of various defects.

In order to improve the most relevant defect, polishing scratch, the modeling on the morphology of the defect was developed and connected to SP1 and the database system, so that the source and mechanism of defect generation are easily and quickly identified and analyzed. The validity of the modeling was further confirmed with the microscopic instruments such as Nomarsky Microscope and AFM with sophisticated data transfer function from SP1 and database system. Several examples of their application during the continuous process improvement activity are reported.

Introduction

Silicon wafer manufacturing for microelectronic industry has to face challenging demand of surface microdefect reduction both in terms of total counting and in terms of detectable size in line with the technological nodes of device integration. At the same time, manufacturing costs are expected to be continuously controlled and reduced.. The combined demand of quality and technology improvement as well as cost reduction can only be managed by a clever application and use of in line process control techniques which assure a fast product quality feedback and process tuning in real time.

Chemical Mechanical Polishing is a key process in determining silicon wafer front surface characteristics (1) among which residual surface scratches are a relevant category. Automatic surface inspection tools based on laser scattered light detection are the instruments used to map in real time the whole surface of all the manufactured wafers. In this way, a huge amount of data related to silicon wafer surface characteristics are daily collected.

The paper will examine examples of surface scratch data analysis combined with polishing process modeling and statistical tools which have provided relevant insight to continuous process improvement and manufacturing cost reduction. The paper will also discuss when an intelligent use of off line microscopic techniques need to be applied.

Experimental /Methodology

Laser Scanning Inspection

KLA Tencor SP1-TBI is a surface inspection system capable of detecting events as small as 60 nm.(2) It is equipped with a 30 mW Ar ion laser and a wavelength of 488 nm.

The laser beam scans the wafer surface in a spiral like pattern and its angle of incidence can be normal (0º) or oblique (70º). The scattered light is collected by two indipendent photomultipliers : Dark Field Wide (DFW) and Dark Field Narrow (DFN) channels. The DFN detects light scattered in an angle from 5º to 20º with respect to the normal at the surface. The DFW collects light scattered in a wider angle between 25º and 70ºC. Both channels are calibrated separately by using polystyrene latex spheres. Measurements performed by the two photomultipliers are then combined in a composite map.

Nomarsky Microscope

Interferometric (Nomarsky) microscope is widely used to inspect polished surface of silicon wafers because its vertical resolution is as high as several tens nm. For the conventional characterization of scratch defect, Model INM200 of Leica was used in a clean room environment. Because the dimension of the defects in subject is too small to be located quickly to the position under objective lens, the microscope is equipped with a Wafer Review Station, in which the coordinates of the defect detected by SP1 are transferred to microscope and the defect is positioned automatically into the view field.

Atomic Force Microscope (AFM)

AFM, Digital Instrument Dimension 5000 was used for further characterization of scratch defect after Nomarsky observation (3). The major parameters of the measurement are the following: curvature radius of silicon tip 5-10 nm, spring constant 20-100 N/m, resonance frequency 200-400 Hz. Several scan sizes from 80μmx80um to 2μmx2μm were applied. Like in case of Nomarsky Microscope, the coordinates of the defects detected by SP1 are transferred to the AFM System.

Manufacturing Database System

Given the huge volume of daily production of silicon wafers, which are in turn divided into small groups by the various specific requests from customers, automatic data control is indispensable for the management and assurance of the quality. Through polishing to final laser inspection, all wafers have their identification codes and are tracked and controlled by Computer Integrated Manufacturing System. The results of SP1 inspection with two-dimensional defect maps (Tencor File Format, TFF maps) are stored in the system combined with all process history including the process parameters and other measurement results. For the rejected wafers, the reason is assigned as defect category such as LPD count by size, area defect, scratch, or haze.

Material

Data reported in this paper are related to polished silicon wafers of 200 mm in size, <100> oriented in a wide range of resistivity and dopant type.

Preliminary Manufacturing Data Analysis

<u>Sub-categrization of Scratch Defects on TFF Maps</u>

Scratch images were observed on TFF maps stored in the database system in order to identify the morphology of the defects. All scratch defects were categorized into the following sub-categories according to their appearance or possible causes. Figure 1 shows typical TFF maps for each sub-categories.

i) <u>Polishing Scratch</u> The smooth and narrow line with the curvature of semi-circle.

ii) <u>Short Scratch</u> The smooth and narrow line but too short to be identified as polishing scratch. Preliminary study under microscope indicates its major part is the short version of polishing scratch.

iii) <u>Handling Scratch</u> Irregular line defect with visible width. It is considered to be damage by some contact of other materials onto the surface caused by the mechanical or handling problem on processing machines.

iv) <u>Contamination</u> Some area defect caused by the contamination is judged as scratch by SP1 algorithm when it finds linear pattern inside the defect.

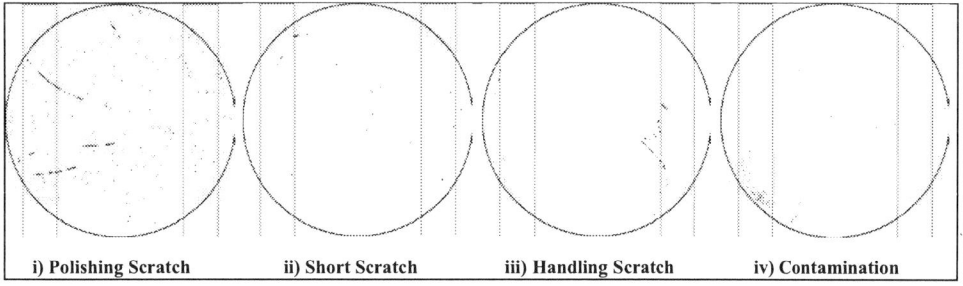

Figure 1. Example of TFF Maps for Sub-categories of Scratch Defects

Figure 2 below summarizes the breakdown of all scratch defects sampled.

Assuming that major part of short scratch is also short polishing scratch, it is concluded that more than a half the scratch defects in the sample wafers are polishing scratch, in other words, polishing process-origin. Therefore, further investigation on polishing scratches was initiated as reported in the following section.

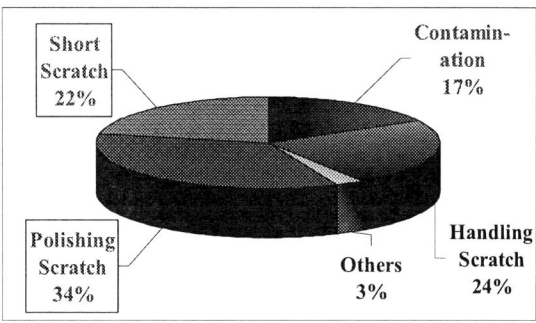

Figure 2. Scratch Loss Breakdown

Polishing Scratch Modeling

Simulation of Polishing Scratch

An own-developed software simulator of the polishing process was used to describe point trjectories , thus allowing a fast and automated simulation of scratch occurrence with related geometrical data. .

As shown in Figure 3, polishing scratch is supposed to be generated when a small but hard particle is fixed on the polishing pad which is pasted on the rotating turntable.

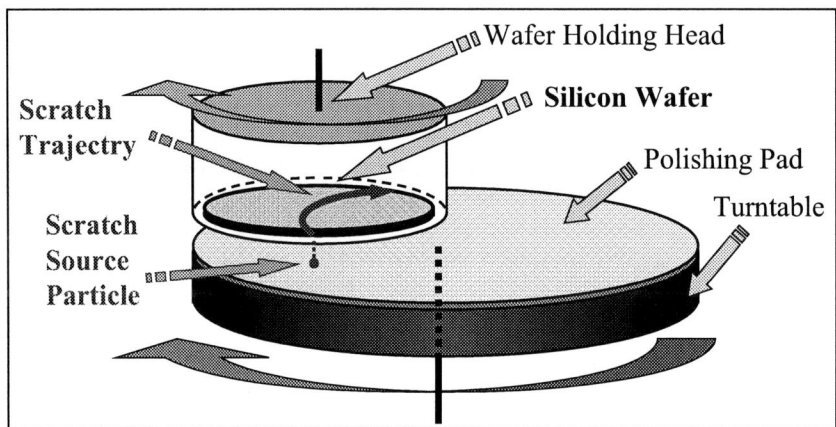

Figure 3. Configuration of Polisher Unit and Scratch Trajectory

By inputting all the dimensional parameters of the machine and the movement parameters, rotation rates of turntable and wafer as well as oscillation of the head, the curves of polishing scratch is reproduced.

Figure 4 shows several examples of simulation result for various rotation speeds of wafer with fixed turntable speed. The simulated curves match very well the observed scratches on TFF maps in terms of curvature and position.

General conclusion is, as wafer rotation speed increases, the scratch becomes more curved.

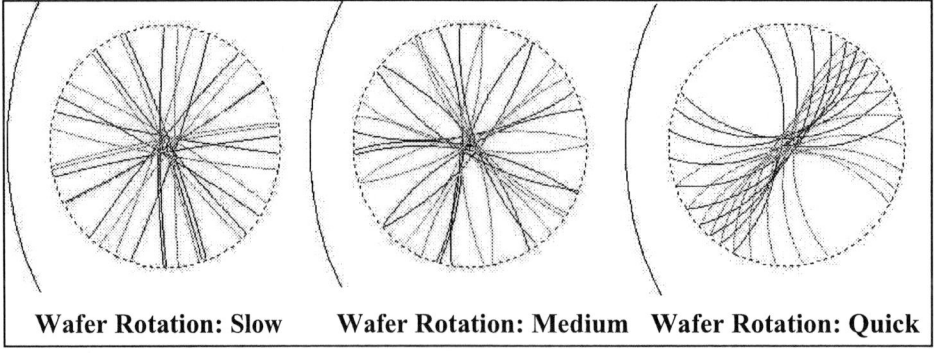

Figure 4. Scratch Trajectory Simulated with Various Rotation Speeds of Wafer

Polishing processes generally consists of multiple polishing phases. The first phases are designed to achieve better flatness for which the rotation speed of wafer can be adjusted depending on the geometric condition of pad. On the other hand, the last phase is designed to obtain finished surface as smooth and clean as possible. In each phase the rotation speed can change and, as a result, scratch curvature will differ phase by phase. In other words, the generation source of scratch can be identified by measuring the curvature and comparing it with rotation speed of each phase of the process used for the subjected wafer.

For this purpose, further quantitative study was carried on by the simulator.

Figure 5 and 6 show the simulated radius distribution of scratch curvature (radius) by various wafer rotations.

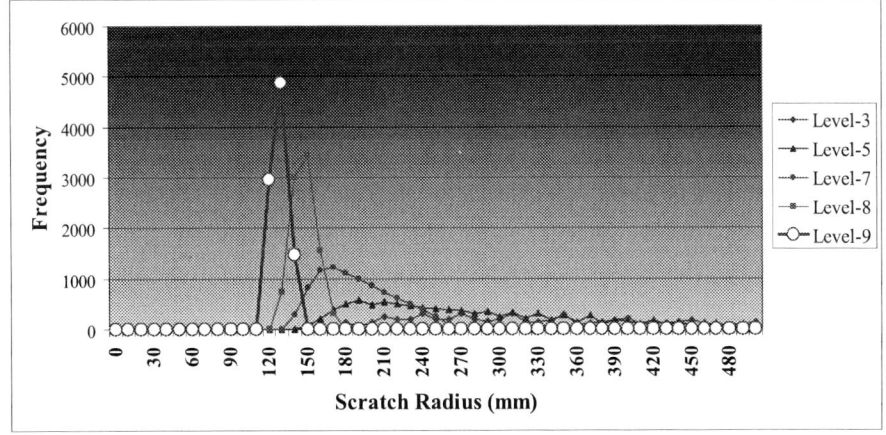

Figure 5. Scratch Radius Distribution vs. Wafers Rotation Speed
(Simulated by Higher Speed, Level-9 : Highest)

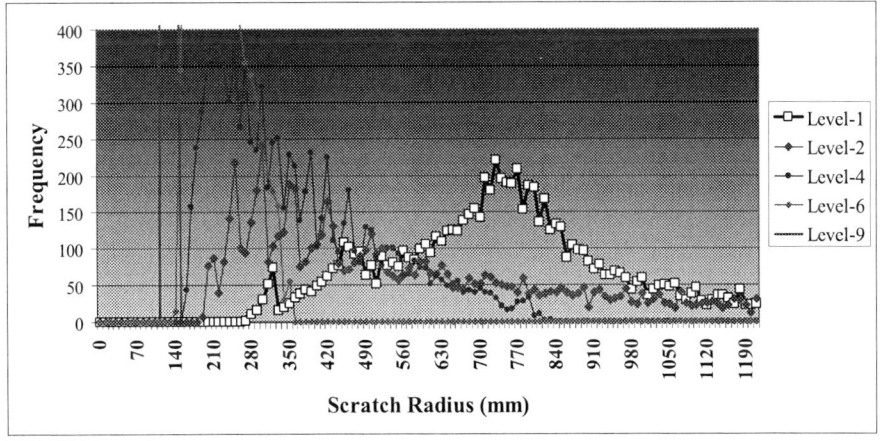

Figure 6. Scratch Radius Distribution vs. Wafers Rotation Speed
(Simulated by Lower Speed, Level-1 : Lowest)

The quantitative data permit to judge the source of polishing scratch with reliable significance. For example, when the first phase rotates wafer around level 1 and the last phase in around level 9, the scratch with radius of 130mm is surely generated in the last phase of the process.

Measurement of Radius on TFF Maps

For full utilization of simulation result, the precise measurement of scratch radius on the map is indispensable. We developed the automatic function of radius calculation on TFF maps inside of database system. The portion of scratch area is identified by dragging the mouse and all points that are part of a scratch are used for the calculation with an adapted least square curve fitting. After radius calculation, the system automatically refers the wafer rotation speed of each polishing phase then judges the source .

In the past, the only way for differentiation of scratch source was microscope observation. Beside its qualitative output, this type of off-line investigation has serious disadvantage that sample size is quite limited respecting to the huge volume of the production. The methodology developed here permits to handle the statistically significant size of data to understand the mechanism of the defect generation.

Result of Scratch Source Identification

Figure 7 shows the result of scratch source identification by the radius analysis on the polishing process with three polishing phases (so called rough, medium, and final from up-flow). Data were collected from three different groups of machines (Line 1, 2, and 3).

Data clearly indicate that major contributor is final phase, especially in Line 3 which performed worst in terms of scratch losses among three lines. The investigation on Line 3 and its improvement action will be described in the last section.

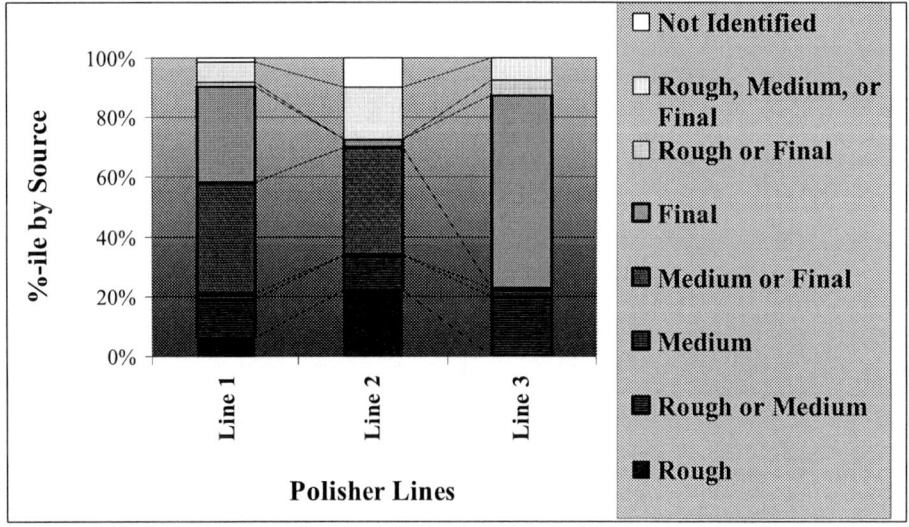

Figure 7. Polishing Scratch Source Distribution

Off-line Characterization for Confirmation of Modeling

For further confirmation of modeling and radius analysis validity, some wafers analyzed for scratch radius were subjected to the observation under Nomarsky microscope and AFM.

Nomarsky Microscope Figure 8 shows the typical microscope images of scratch from various sources which were judged by radius analysis on TFF maps. Rough-origin scratch is characterized by the deep point damages connected by line damage. Medium-origin scratch is either deep line or line-up of point damages. In both cases, the damages are evidently polished off, in other words, the boundary is smoothed off by subsequent polishing. On the contrary, final-origin scratch is very thin and shallow with no evidence of polish-off.

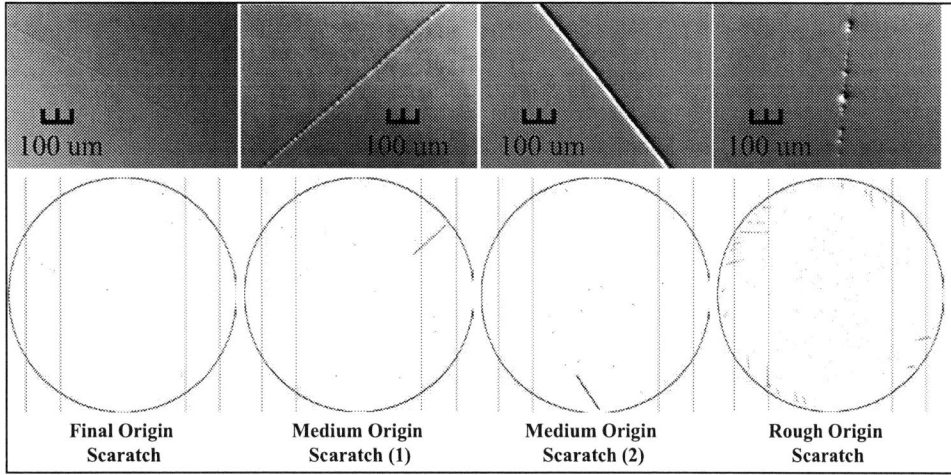

Figure 8. Nomarsky Microscope Images of Polishing Scratch from Various Sources

Different morphology of scratches among three sources is explained by the difference of stock removal among each steps. The task of rough polisher is to create mirror surface and to define the required flatness so the removal is highest. The task of medium is fine adjustment of the flatness so it removes the silicon few times less than rough. That of final is to finish the surface with finest slurry grits so the removal is almost negligible.

Even if the scratch source with small size exists on rough or medium polisher, the generated scratch will be removed off afterward thanks to enough stock removal. On the other hand, the same small particles generate scratches at final, but all of them will survive because of almost no removal after it. Only scratches generated by fewer but larger sources at up-flow survive with the trace of polish-off. This model explains also why final-origin scratch is the majority of polishing scratch. Final polisher is the most vulnerable for scratches because of no recovery after that.

AFM Figure 9 shows the example of AFM 3D images with 2μmX2μm scanning taken from rough-medium origin and final-origin polishing scratches.

Rough-medium origin scratches contain the pits without exception, indicating that the surface was previously damaged and later polished off, while the final-origin shows continuously smooth and shallow line, indicating that it is relatively fresh.

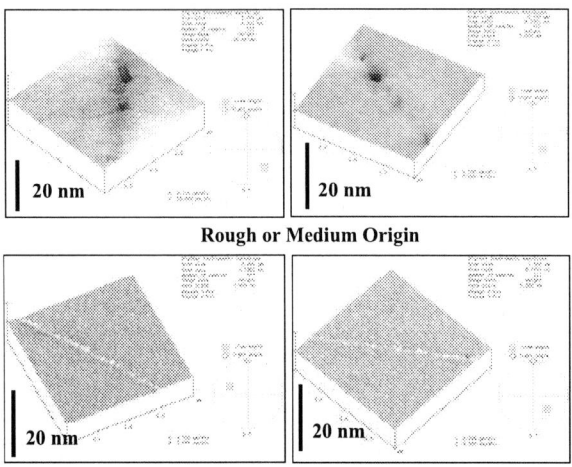

Rough or Medium Origin

Final Origin

Figure 9. AFM 3 D Images of Rough or Medium Origin and Final Origin Scratches

Improvement Examples

Hereafter, a couple of examples of process improvement are reported based on the information obtained by our methodology consisting in   SP1 inspection, database system, curvature analysis of scratch, and microscope and AFM observation.

Final-phase Polishing Scratch Reduction

Knowing that major part of scratch reject is polishing scratch of which final-origin scratch is dominant, the difference in final polishing between the best performing line (Line 1) and other two groups were investigated. After aligning the other two groups like as Line 1 for small differences in the equipment hardware of the slurry distribution and pad cleaning, total scratch defect level was down as shown in Figure 10.

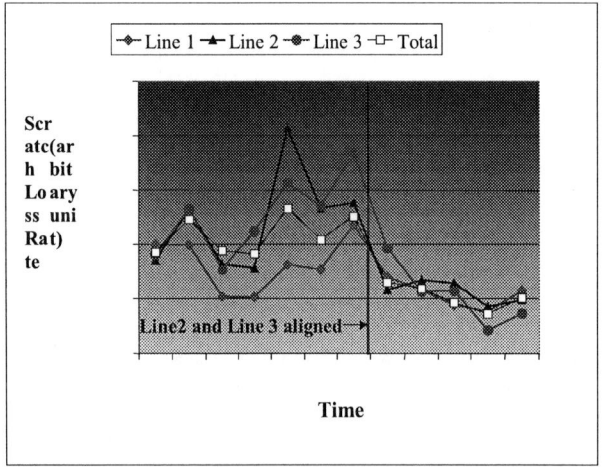

Figure 10. Scratch Loss Trend by Polishing Lines

Medium-phase Polishing Scratch Reduction

Rough and medium-origin scratch was found to show up intermittently, unlike final-origin which was uniform on time like a background. By commonality analysis it was found to be caused by wrong re-conditioning of aged pads. As shown in Figure 11, the upset of rough and medium-origin scratches almost disappeared after establishing the proper procedure for this operation.

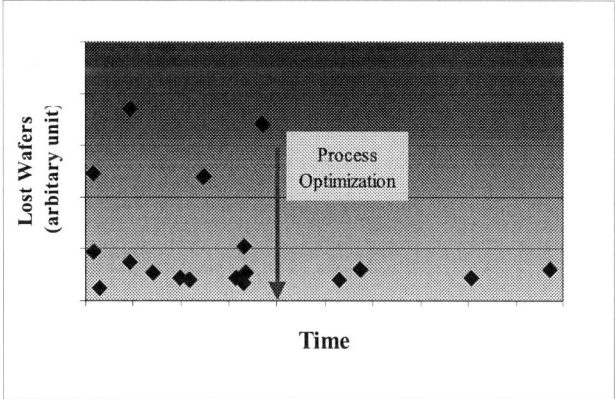

Figure 11. Improvement of Rough and Medium Scratch

Conclusions

The paper has discussed a systematic methodology of in-line surface defect control during silicon wafer polishing process. Examples have been provided to show the relevance of an intelligent combination of data analysis and modeling with microscopic observation of the defects in order to control and improve in real time a manufacturing process.

References

1. J.J.Shen, L.M.Cook, K.G.Pierce, S.B.Lonckl, J.Electrochem. Soc.,143 (6), p.2068 (1996)
2. G.Lorenzi,K.H.Nguyen,C,Sanna,R.Orizio, G.Borioentti"COPs/Particle discrimination using an automated surface inspection tool" SPIE's International Symposium Advanced Microelectronic Manufacturing Santa Clara CA Feb 2003
3. G.Borionetti, A.Bazzali,R.Orizio , Eur.Phys.J.Appl.Phys., 37, p.101 (2004)

Molybdenum Contamination in Indium and Boron Implantation Processes

D. Codegoni, M.L. Polignano, V.Soncini, C. Bresolin

ST Microelectronics, Via Olivetti, 20, 20041 Agrate Brianza (MI) Italy

> Some examples of contamination by mass interference are studied
> in this work, specifically the molybdenum contamination in indium
> and boron implantations. This contamination is detected by Elymat
> and DLTS analyses, whereas it is missed by SPV, which is
> commonly used for routine line monitoring. Both Elymat and
> DLTS analyses consistently show molybdenum contamination
> clearly related to indium implantation conditions and to high dose
> boron implantations.

Introduction

It is known (1) that ion implantation can be responsible for serious transition metal
contamination because of various mechanisms, among which sputtering was found to be
dominant in the case of high dose implantations with heavy ions (2). Transition metal
contamination by sputtering was shown to be essentially an iron contamination, and
methods to reduce this sort of contamination were identified. In the evolution of the
implantation technology this mechanism was strongly reduced, so that it is no longer the
dominant contamination mechanism, at least in new equipment (3). In low dose
implantations contamination by sputtering is not so important because this mechanism is
responsible for a contaminant concentration that increases in proportion to the implanted
dose. However, low dose implantations can in specific cases be affected by
contamination problems because of mass interference. In this work an example of this
sort was studied, namely the molybdenum contamination by mass interference. This
contamination may affect indium implantations in the presence of fluorine, as the
molybdenum-fluorine complex has almost the same mass as indium. Boron implantations
too may result in molybdenum contamination by mass interference because of the
formation of multiply ionized molybdenum ions and complexes (4,5).

Contaminants introduced by sputtering usually receive a few keV only, and hence
penetrate into the silicon matrix by less than 100 Å. Therefore, the silicon contamination
due to this mechanism can be strongly reduced by a screen oxide, and on the other hand it
can be detected by Total Reflection X-ray Fluorescence (TXRF). On the contrary,
contamination by mass interference involves energetic contaminant ions, which penetrate
deep into silicon and are hence frequently missed by TXRF. This contamination can be
only detected by techniques able to reveal low concentration elements in the silicon
volume. For this reason, in this study we used techniques based upon the properties of
molybdenum as a deep level in silicon, namely carrier lifetime measurements and the
Deep Level Transient Spectroscopy (DLTS). Carrier Lifetime was obtained from
enhanced Surface Photovoltage (SPV) measurements and from photocurrent
measurements in the Elymat (Electrolytic Metal Tracer) technique.

Experimental Details

Sample preparation

Molybdenum is a very well know slow diffuser contaminant (6), and it has a deep level located about 0.3 eV above valence band and hence acting a hole trap. In a previous work about molybdenum as a silicon contaminant (7), we observed that molybdenum is prone to surface segregation during furnace treatments with a slow temperature ramp-down, so the amount of molybdenum in the solid solution in silicon is reduced when these treatments are used. These features impose some constraint to the flow to be followed for the sample preparation. First, p-type material has to be used to reveal molybdenum by DLTS as a majority carrier trap. Therefore, p-type, 725 μm thick, Magnetic Czochralski (MCZ) grown 8 inch wafers with 10 Ωcm resistivity were used for this study. The MCZ material was chosen to reduce the risk that oxygen precipitation nuclei may confuse lifetime data.

Molybdenum contamination may affect indium implantations in the case the formation of a molybdenum-fluorine complex is possible, because this complex has almost the same mass as indium. Therefore, implantation conditions are assumed to be critical if molybdenum and fluorine are simultaneously present before the mass selection. Some wafers were implanted with indium (150 keV, $8 \cdot 10^{13}$ cm^{-2}) under various conditions ranging from the most critical to the less critical for molybdenum contamination according to this criterion (see table I). A medium current, single wafer implanter equipped with a molybdenum source (Varian E500) was used in this experiment. To avoid molybdenum surface segregation, a rapid thermal treatment (1100°C, 200 s) was chosen to diffuse the contaminant after the implantation, though this treatment is expected to diffuse molybdenum through a limited portion of the wafer only. Then, a 1 μm wet silicon etching was used to remove the indium-doped layer after contaminant diffusion. This step cannot be avoided, even if some contaminant is removed along with doping. Indeed, the doped layer must to be removed before measurements, because doping itself has an impact on carrier lifetime. In addition, DLTS measurements require the formation of a Schottky contact, which would not be possible with such a high doping. The samples for lifetime measurements received no further treatment, in the samples for DLTS measurements Schottky diodes were created by depositing, masking and etching a titanium layer. The latter operations were carried out by limiting sample heating as much as possible, in order to prevent contaminant surface segregation. Some not implanted samples were also prepared according to this flow for a reference.

TABLE I. Presence of fluorine compounds before indium implanation

Non-critical conditions	Critical conditions	Most critical conditions
No	BF₃ for 2h 20 min	BF₃ for 7h 15 min

An experiment about high dose boron implantations ($1.6 \cdot 10^{15}$ cm^{-2} and $4 \cdot 10^{15}$ cm^{-2}, 60 keV) was also carried out. A batch, high current implanter with a molybdenum source (Axcelis GSDIII/LED) was used. After the implantation, the wafers followed the same flow as used for the indium implantation experiment.

Experimental techniques

DLTS. 0.785 mm^2 area Schottky diodes were used for DLTS measurements. A DLS-83D instrument by Semilab was used. In this instrument lock-in integration is used for averaging capacitance transients, and temperature can be scanned from about 30 K to 300 K. Alternatively, constant temperature spectra can be obtained as a function of the frequency of excitation pulses (8) in the range 0.5 Hz–2 KHz. Both methods were used in this work. The differential DLTS method (9) was used. In this method the Schottky diode (or the p-n junction) is reverse biased at a voltage V_r and two filling pulses are applied, a first pulse V_1 in the beginning of the lock-in integrating period and a second pulse V_2 a half period later. In the lock-in integration, the difference ΔC is obtained between the integrals of the capacitance transients caused by the first pulse and by the second pulse. The differential DLTS can also be used to obtain the in-depth trap concentration profile. Indeed, this method yields the trap concentration in the interval $(x_d(V_1), x_d(V_2))$, (where x_d is the depletion region edge at a given reverse voltage) so by suitably choosing V_1, V_2 and V_r the trap concentration can be measured as a function of depth.

In our measurements, samples were reverse biased at -5V, and filling pulses with 4.5 V and 0.5 V amplitudes were applied during each integrating period with a pulse width of 10μs. When measuring the trap concentration profile, V_1- V_2 was set at 1V and V_r was run up to 20 V, thus allowing about 3 μm depth to be explored.

When comparing DLTS and lifetime data, it is necessary to recall that DLTS measurements explore the space charge region of the junction only, which is of the order of a few μm, i.e. much closer to the surface than the region probed by any recombination lifetime technique.

Elymat. In this technique excess carriers are generated by a laser beam, and the diffusion length is extracted from photocurrent measurements (10). The contribution of surface recombination is suppressed by immersion in a HF solution (11) which also provides Schottky contacts for carrier collection. Carrier diffusion length (or carrier lifetime) data are obtained by injecting carriers at the wafers frontside and measuring the photocurrent at the wafer backside (backside photocurrent, BPC) or at the wafer frontside (frontside photocurrent, FPC). In this work lifetime measurement were obtained from BPC measurements by injecting carriers with an 830 nm wavelength laser.

The Elymat technique allows the injection level to be varied by varying the intensity of the laser beam, so that measurements of lifetime as a function of the injection level are possible. In p-type silicon, the injection level is defined as $\delta n/p_0$, where δn is the minority carrier excess and p_0 is the equilibrium majority carrier concentration. In previous works (12) it was shown that data of lifetime vs. injection level can be used for the identification of contaminants. Indeed, according to the Shockley-Read-Hall theory, the dependence of recombination lifetime τ on the injection level is uniquely determined by the properties (energy levels and capture cross sections) of the dominant recombination centers. If one impurity dominates recombination in a sample, the function $\tau(\delta n/p_0)$ is a feature of the corresponding recombination centers, and this allows us to identify this impurity. A quantitative estimate of its concentration can be obtained by a best-fit of the theoretical $\tau(\delta n/p_0)$ curve to the experimental data with one fitting parameter only, i.e. the impurity concentration. This method was validated for the quantitative evaluation of various contaminants, e.g. iron, chromium, etc., as well as molybdenum, by using metal-implanted wafers. Iron in p-type silicon can be easily identified by the formation and dissociation of iron-boron pairs. Iron is under the form of FeB pairs at equilibrium, however FeB pairs can be dissociated by baking the samples (180°, 1 h baking was used in this work), and immediately after dissociation interstitial iron dominates carrier recombination. Iron in p-type silicon has a characteristic fingerprint in that FeB pairs and interstitial iron have opposite lifetime dependence on the injection level, namely the carrier lifetime decreases with increasing the injection level when it is limited by FeB pairs, and increases when it is limited by interstitial iron. Vice versa, no lifetime changes were detected by baking molybdenum contaminated samples (7).

In backside photocurrent measurements, in principle the whole wafer thickness is probed. However, in the characterization of molybdenum as a recombination center (8) it was observed that BPC Elymat data are sensitive to the in-depth distribution of recombination centers, possibly because lifetime data are somehow weighted by the excess carrier concentration. In the case of slow diffusers, the in-depth distribution is likely to be non-uniform, so comparing Elymat measurements taken at wafer frontside and at wafer backside can be an indication of the presence of this sort of contaminants.

If the surface recombination is not suppressed by immersion in a HF solution, but rather determined by an oxide-silicon interface or by the presence of metal precipitates at the wafer surface, the photocurrent collected at wafer backside is affected by both bulk and surface recombination. In this case, it is possible to obtain both the bulk recombination lifetime and the surface recombination velocity by subsequent measurements of the backside photocurrent (13,14).

SPV. Surface photovoltage (SPV) measurements are carried out by illuminating the sample with light of various wavelengths. Generated minority carriers are collected in a depletion region at the wafer surface and produce a variation in surface potential, which is recorded as a function of light wavelength. In the standard SPV technique (15), the carrier diffusion length L_{diff} is extracted from these data by assuming that sample thickness is much larger than the diffusion length. This hypothesis often fails in present wafers, and the actual wafer thickness must be taken into account. An "enhanced SPV" (16) technique is available for this purpose. The enhanced SPV equation is obtained by imposing the correct boundary condition at the wafer backside, i.e. by taking into account backside surface recombination. The surface photovoltage signal ΔV is written as $\Delta V \propto f(z, L_{diff})$, where f is a function of the light penetration depth z and of the carrier diffusion length. The carrier diffusion length is obtained by a non-linear best-fit procedure to surface photovoltage data. The surface recombination velocity is included in the proportionality constant and is eliminated by normalization to the shortest wavelength datum, so the obtained carrier diffusion length is affected by bulk recombination only. SPV measurements are always obtained under very low injection conditions.

In SPV measurements the probed region is of the order of light penetration depth plus carrier diffusion length, so in these experiments it is of the order of the whole wafer thickness. Even in these conditions, in the case of a non-uniform in-depth distribution of contaminants SPV diffusion length measurements were found to be more sensitive when contaminants are located near the surface exposed to the light source (17), similarly to what observed for Elymat BPC data.

In this work the SPV tool in the FAaST SDI system was used. This instrument is equipped with a set of filters producing monochromatic light at seven different wavelengths, ranging from 800 nm to 1 μm. The enhanced SPV method was used.

SPV measurements allow iron to be identified in p-type silicon by optical dissociation of the iron-boron pair (16).

Experimental Results

Indium implantation experiment

SPV results. Table II collects the SPV results. SPV measurements do not show a significant concentration of any other contaminant, apart from a limited iron concentration. Whether the contamination is due to sputtering or to mass interference, contamination by implantation comes from wafer frontside, so a lower diffusion length at wafer frontside than at wafer backside indicates a not complete redistribution of the contaminant. Molybdenum is a slow diffuser, so the frontside-to-backside difference is expected to be much more significant in the presence of molybdenum. The diffusion length data measured at the frontside of implanted wafers are lower than the data measured at wafer backside, but this difference is always less than the spread over the wafer and in addition it does not show any trend with implantation conditions. Concluding, SPV did not detect any relevant impact of the implantation conditions on carrier diffusion length and on contaminant concentration.

TABLE II. SPV measurements of indium-implanted samples.

Implantation conditions	L_{diff} (μm)		[Fe] (10^{11}cm^{-3})	Other impurity concentration (10^{10}cm^{-3})
	frontside	backside		
No(reference)	770 ± 170	650 ± 170	3.6 ± 1.6	1.3 ± 1
Non-critical	500 ± 85	590 ± 120	3.1 ± 1.4	1.6 ± 0.9
Critical	490 ± 90	590 ± 140	3.2 ± 1.4	1.7 ± 1
Most critical	620 ± 170	630 ± 130	3.1 ± 1.4	1.6 ± 0.8

Elymat results. Table III reports the Elymat measurements of carrier diffusion length obtained with 1 mA injected current. In this case some trend with implantation can be recognized, though the effect is limited. However, a clear feature related to indium implantation conditions can be identified by measuring carrier lifetime as a function of the injection level. Fig. 1 reports the carrier lifetime as a function of the injection level in a reference sample (a) and in samples implanted under different conditions ((b), non-critical, (c), critical, (d), most critical). The measurements were repeated before and after baking. For comparison purposes, fig. 2(a) reports the carrier lifetime vs. injection level in an iron-implanted sample before and after baking, and fig. 2(b) reports the carrier lifetime vs. injection level in samples implanted with two different molybdenum doses. In figs. 1(a—b)) and 2(a) the lines are the curves obtained by the best-fit procedure which also yields an estimate of the iron concentration. By comparing these data, it is immediately recognized that the in the reference sample (fig. 1(a)) the lifetime is limited by a moderate iron contamination, probably coming from the thermal treatment or from handling. On the opposite side the sample implanted with indium under the most critical conditions (fig. 1(d)) clearly shows the presence of another contaminant, that does not change its activity upon baking. The samples implanted under non-critical and critical conditions (fig1 (b) and (c)) show a gradual departure from the iron-limited lifetime behaviour. This fact is especially evident in the curves taken before baking the samples, when iron is under the form of FeB pairs, which correspond to a very specific dependence of the lifetime on the injection level. The same iron concentration is estimated from the data in fig. 1(a—c), confirming that iron contamination is not related to the implantation conditions. The carrier lifetime dependence on the injection level in fig. 1(d) is similar to what observed in molybdenum-implanted wafers. However, this dependence is not unique and DLTS measurements are required to confirm molybdenum contamination. It is also worth recalling that the data in fig. 2(b) were obtained in furnace annealed wafers, so that a significant molybdenum amount segregated at wafer surface. This phenomenon was responsible for the lifetime increase with increasing molybdenum dose, opposite to what is expected when all the contaminant is in the solid solution in silicon.

TABLE III. Carrier diffusion length of indium-implanted wafers, as measured by Elymat with 1mA injected current.

Implantation conditions	$L_{diff} (\mu m)$
No (reference)	690 ± 80
Non-critical	600 ± 40
Critical	575 ± 30
Most critical	510 ± 20

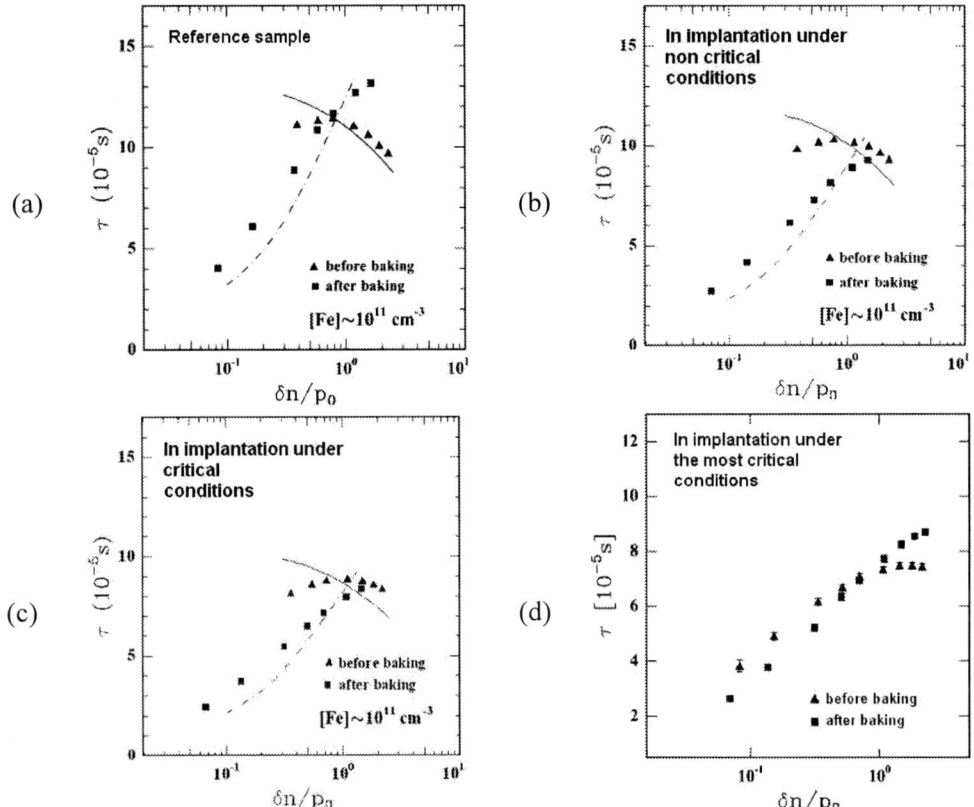

Figure 1. Carrier lifetime vs. injection level in samples implanted with indium under different conditions (b—d) and in a reference sample (a).

DLTS results. DLTS spectra show a characteristic peak in the sample implanted under critical and most critical conditions, as shown in fig. 3. Fig. 4 reports the Arrhenius plot of e_p/T^2 (where e_p is the hole emission rate and T the absolute peak temperature) obtained from the DLTS spectra of indium-implanted samples. Molybdenum literature data are also reported for a comparison (6) and allow the peaks in fig. 3 to be identified with molybdenum.

In-depth concentration profiles were obtained for the peak in fig. 3. The molybdenum concentration in the solid solution was found to be approximately constant in the range explored by this technique (about 3μm depth, see fig. 5), apart from some depletion in the region close to the surface, possibly due to surface segregation. As previously mentioned, molybdenum surface segregation was observed in experiments using furnace treatments for molybdenum diffusion (7). In the present experiments using RTP treatments this phenomenon is reduced but maybe not completely suppressed.

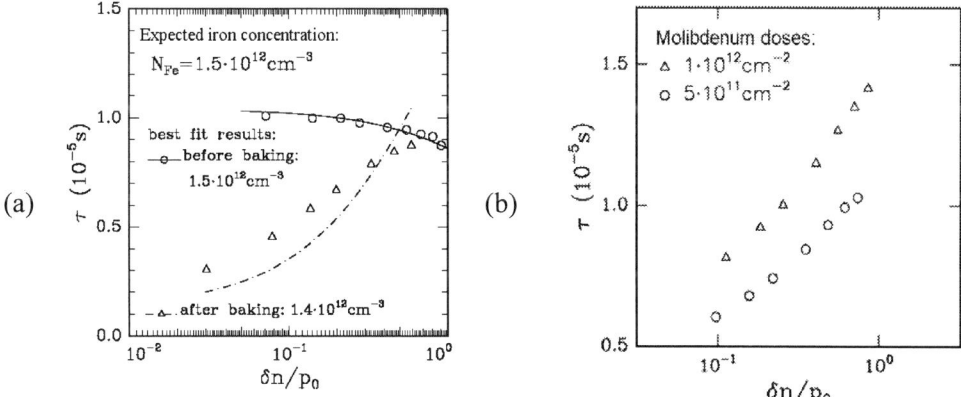

Figure 2. Carrier lifetime vs. injection level in a sample intentionally contaminated by implantation with iron (a) and with molybdenum (b)

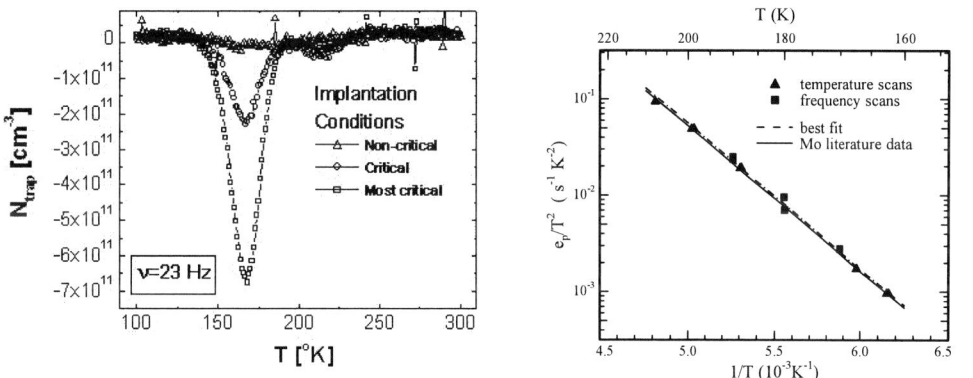

Fig. 3. DLTS spectra of samples implanted with indium under various conditions.

Fig. 4. Arrhenius plot of e_p/T^2 obtained from the peaks in fig. 3.

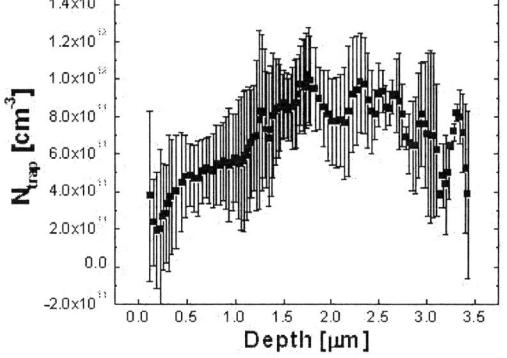

Fig. 5. Molybdenum concentration profile in a sample implanted with indium under the most critical conditions, as obtained by differential DLTS.

Boron implantation experiment

SPV results. Table IV compares the SPV diffusion length measured at wafer front-side and at wafer backside of wafers implanted with a $1.6 \cdot 10^{15} \text{cm}^{-2}$ boron dose and of a reference wafers receiving the RTP treatment only. In both samples, the SPV diffusion length is lower at wafer backside than at wafer front-side, due to a moderate iron contamination. No effect of the boron implantation is detected.

TABLE IV. SPV diffusion length of boron-implanted wafers and of reference wafers.

Boron dose	L_{diff} (µm)		[Fe] (10^{10}cm^{-3})	Other impurity concentration (10^{10}cm^{-3})
	frontside	backside		
No(reference)	800± 140	590 ± 60	0.4 ± 0.3	1.4 ± 0.4
$1.6 \cdot 10^{15} \text{cm}^{-2}$	797 ± 140	560 ± 70	0.8 ± 0.4	1.5 ± 0.5

Elymat results. Table V reports the Elymat diffusion length measured at wafer front-side and at wafer backside of wafers implanted with a $1.6 \cdot 10^{15} \text{cm}^{-2}$ boron dose and of a reference wafers receiving the RTP treatment only. In this case, a significant effect of the boron implantation is observed in the measurements taken at wafer front-side. The diffusion length measured at wafer backside is not affected by the boron implantation, indicating that the degradation measured at wafer front-side is due to a slow diffuser.

TABLE IV. Elymat diffusion length of boron-implanted wafers and of reference wafers.

Boron dose	L_{diff} (µm)	
	frontside	backside
No(reference)	1260± 380	1120 ± 340
$1.6 \cdot 10^{15} \text{cm}^{-2}$	474 ± 30	1202 ± 420

Fig. 6 reports the carrier lifetime vs. injection level in samples implanted with $1.6 \cdot 10^{15} \text{cm}^{-2}$ and with $4 \cdot 10^{15} \text{cm}^{-2}$ boron doses. The observed dependence is similar to what shown in figs. 1(d) and 2(b), suggesting that molybdenum is responsible of the observed lifetime degradation. DLTS measurements were carried out to confirm this indication.

DLTS results. Fig. 7 shows the DLTS spectrum of a sample implanted with $4 \cdot 10^{15}$ cm^{-2} boron dose, confirming the presence of molybdenum in this sample. This spectrum also shows a low concentration peak at about 220 K, corresponding to a level located at about 0.4 eV above the valence band edge. This peak is ascribed to an iron-related level [18] different from the FeB pair, and can also be observed by a careful inspection of fig. 3. The FeB level located about 0.1 eV above the valence band [19] was not observed in these analyses.

Discussion and Conclusions

The previously discussed experiments show two examples of molybdenum contamination induced by ion implantation processes because of mass interference phenomena. Our results show that this contamination is not detected by SPV if a RTP treatment is used to diffuse contaminants into the bulk. However, in a previous work we showed that molybdenum is revealed by SPV [7] if long furnace treatments are used to diffuse it, so we suggest that in the present experiments molybdenum is missed by SPV because it is too close to the wafer surface. Previous findings (7,17) indicate that SPV is sensitive to contaminants located near the surface where carriers are injected, however in those experiments the estimated deep level profile extended through a few hundred micron. Here, we expect that the contaminant distribution is much closer to the wafer surface, though no estimate is possible due to the lack of diffusivity data. The Elymat technique detects molybdenum irrespective of its in-depth distribution. Actually, we recall that the Elymat technique is sensitive to the surface recombination velocity (13,14), and a recombination center distribution close to the wafer surface can be similar to a surface recombination if compared to the probing depth of the BPC Elymat configuration (the whole wafer thickness). In any case, it is important to take into account that this sort of contaminants may escape a SPV analysis, which is commonly used for line monitoring.

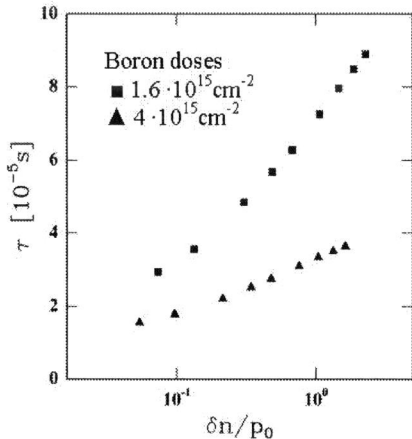 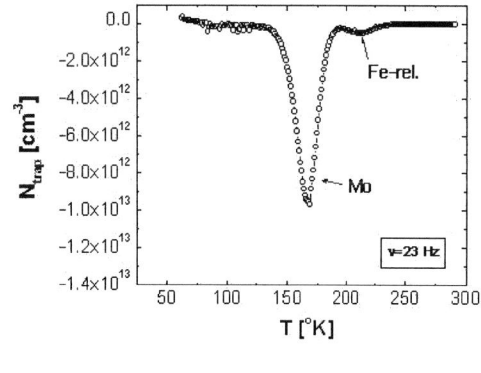

Fig.6. Carrier lifetime vs. injection level in high dose boron-implanted samples.

Fig. 7. DLTS spectrum of a sample implanted with $4 \cdot 10^{15} cm^{-2}$ boron dose.

We also note that in both experiments the moderate iron contamination revealed by SPV corresponds in the DLTS analysis to an iron-related peak different from the FeB pair peak. The iron-related peak at 0.4 eV above valence band was also observed in a DLTS analysis of arsenic-implanted samples (20), and also in that case FeB pairs were not revealed by DLTS, though a non-negligible concentration of FeB pairs was measured by SPV. It is therefore suggested that the excess point defects produced by ion implantation enhance the formation of the 0.4 eV level in the near-surface region, which is explored by DLTS, whereas FeB pairs are left in the bulk.

Acknowledgments

This work was supported by the European Commission - Research Infrastructure Action, under the FP6-Program "Structuring the European Research Area", through the Integrated Infrastructure Initiative "European Integrated Activity of Excellence and Networking for Nano and Micro-Electronics Analysis", contract n. 026134 (RII3) ANNA. The authors wish to thank P. Godio (MEMC, Novara, Italy) for giving them access the DLTS system in Novara for some of the DLTS measurements in this work. Thanks are also due to A. Danel (LETI, Grenoble, France) for an interesting discussion about the interpretation of lifetime data.

References

1. J. F. Ziegler, *Handbook of Ion Implantation Technology*, pp.675—691, North Holland, Amsterdam, (1992)
2. M. L. Polignano, C. Bresolin, F. Cazzaniga, A. Sabbadini and G. Queirolo, in *Optical Characterization Techniques for High-Performance Microelectronic Device Manufacturing II*, J.K. Lowell, R.T.Chen, J.P.Mathur, eds., Proc. SPIE 2638, p. 14, SPIE-Int. Soc. Opt. Eng. (1995)
3. M. L. Polignano, D. Caputo, A. Giussani, V. Soncini, G. Di Toma, in *2000 International Conference on Ion Implantation Technology Proceedings*, H. Ryssel, L. Frey, J. Gyulai and H. Glawischnig, eds., p. 686, IEEE, Piscataway, NJ (2000)
4. K. Funk, V. Haublein, H. Chakor, M. Ameen, L. Frey, H. Ryssel, A. Ramirez, *2000 International Conference on Ion Implantation Technology Proceedings*, H. Ryssel, L. Frey, J. Gyulai and H. Glawischnig, eds., p. 711, IEEE, Piscataway, NJ (2000)
5. V. Haublein, L. Frey, H. Ryssel, in *2002 International Conference on Ion Implantation Technology Proceedings*, B. Brown, T. L. Alford, M. Nastasi and M. C. Vella, eds., p. 217, IEEE, Piscataway, NJ (2002)
6. A. Rohatgi, R. H. Hopkins, J. R. Davis, R. B. Campell and H. C. Mollenkopf, *Solid-State Electron.* **23**, 1185 (1980)
7. M. L. M. L. Polignano, C. Bresolin, G. Pavia, V. Soncini, F. Zanderigo, G. Queirolo, M. Didio, *Mat. Sci. Eng. B* **53**, 300 (1998)
8. G. Ferenczi, J. Boda and T. Pavelka, *Phys. Stat. Sol.* (a) **94**, K119 (1986)
9. G. Ferenczi, C. A. Londos, T. Pavelka, and M. Somogyi, A. Mertend, *J. Appl. Phys.* **63**, 183 (1988)
10. V. Lehmann and H. Föll, *J. Electrochem. Soc.* **135**, 2831 (1988)
11. E. Yablonovitch, D. L. Allara, C. C. Chang, T. Gmitter and T. B. Bright, *Phys. Rev. Letters 57*, 249 (1986)
12. M. L. Polignano, F. Cazzaniga, A. Sabbadini, F. Zanderigo, F. Priolo, *Mat. Sci. Eng. B* **55**, 21 (1998)
13. M. L. Polignano, A. P. Caricato, A. Modelli and R. Zonca, *J. of the Electrochem. Soc.* 147 , 76 (2000)
14. M. L. Polignano, F. Cazzaniga, A. Sabbadini, G. Queirolo, A. Cacciato and A. Di Bartolo, *Mat. Sci. Engineering B* **42**, 157, (1996)
15. A. M. Goodman, *J. Appl. Phys.* **32** 2550 (1961)
16. J. Lagowski, P. Edelman, A. M. Kontkiewicz, O. Milic, W. Henley, M. Dexter, L. Jabstrzebski and A. M. Hoff, *Appl. Phys. Lett.* **63** , 3043 (1993)
17. M. L. Polignano, D. Caputo, G. Pavia, F. Zanderigo, *Solid-State Phenomena* **64**, 413-420 (1998)
18. K. Wünstel and P. Wagner, *Solid State Communications* **40**, 797 (1981)
19. S. D. Brotherton, P. Bradley and A. Gill, *J. Appl. Phys.* **57**,1941 (1985).
20. E. Colelli, A. Galbiati, D. Caputo, M. L. Polignano, V. Soncini, G. Salvà, in *Proc. of the 8th International Symposium on Plasma- and Process-Induced Damage, 2003*, K. Eriguchi, S. Krishnan, and T. Hook. eds, p. 81, IEEE, Piscataway, NJ (2003)

CHAPTER 3

TECHNIQUES FOR CHARACTERIZATION OF DEFECTS AND CONTAMINATION

ECS Transactions, 10 (1) 97-108 (2007)
10.1149/1.2773980 ©The Electrochemical Society

Application of selected electron microscopy methods to materials analysis problems

A. Rucki[a] and H. Cerva[a]

[a] Siemens AG, Corporate Technology, Otto-Hahn-Ring 6, D-81730 München, Germany

Materials analysis problems often demand site-specific preparation at sub µm scale and element analysis at nm scale. Focused ion beam instruments and field emission transmission electron microscopes are instruments offering appropriate methods. This contribution features different applications: We show how EDX/EELS investigations help to understand the mechanism and the different stages of worm corrosion encountered during the manufacturing of aluminum interconnects. We present calibration data suitable for the quantitative determination of Ge in SiGe material by EFTEM/ESI at sub 10 nm scale. We describe a 3D defect analysis technique for silicon devices which combines defect localization in thick plan-view samples by 400 kV TEM with subsequent FIB cross-sectioning of the plan-view specimen. We demonstrate the application of FIB tomography to the characterization of delaminations at a chip / molding compound interface.

Introduction

Today focused ion beam instruments (FIB) and field emission gun transmission electron microscopes (FEG-TEM) are indispensable and well-established tools for the characterization of semiconductor materials and devices.

A FEG-TEM offers both energy-filtered imaging (EFTEM/ESI), which is also applicable with a LaB_6 instrument, and the capability of 1nm probe measurements for energy-dispersive x-ray spectroscopy (EDX), and electron energy loss spectroscopy (EELS) for nanometer-scale elemental analysis. The potential of the latter techniques is extended in a FEG-TEM with a scanning unit (STEM) making EDX/EELS mapping and elemental-sensitive Z contrast imaging possible. A modern STEM-EDX/EELS acquisition system both corrects specimen drift during acquisition and stores each acquired spectrum (spectrum imaging) thus allowing mapping at nanometer resolution and the extraction of elemental maps and sum spectra by post-processing of the acquired 3D spectra data cube (1). Contamination under the electron beam, which strongly affects 1nm measurements, can be avoided in many cases by plasma-cleaning of specimen holder and specimen. If this is not sufficient or the material is sensitive to plasma cleaning the use of a cooling holder is a recommended option.

In general the preparation of sample cross-sections by FIB extends the capabilities of the classical grinding and subsequent etching /Argon milling / polishing approach both for single sections in SEM inspection and thin electron transparent lamellae for TEM analysis. Cross-sectional preparations of water-sensitive and organic materials or layer structures with poor adhesion at interfaces are typical FIB applications. TEM lamellae can be prepared by FIB with defined sample thicknesses of 50-100 nm which is optimum for quantitative EDX analysis and thin enough for EELS analysis. Moreover, the capabilities of the focused Ga probe of FIB instruments allow site-specific preparation down to sub 100 nm feature size which is a prerequisite for failure analysis in current semiconductor and magnetic device technologies.

Recent improvements in the field of FIB instruments are the integration of a FIB and a scanning electron microscope (SEM) into a single system, a dual beam FIB, and the development of ion optics which achieve sufficient focusing of the ion beam at accelerating voltages as low as 2 kV. The latter development accounts for the final polishing of the FIB cut and allows a considerable reduction of the FIB induced amorphous surface layer from 10-20 nm to below 2-3 nm, e.g. for silicon (2). This allows high resolution TEM investigations on FIB prepared samples and becomes also relevant for surface sensitive SEM techniques such as low kV imaging and electron back scattered diffraction (EBSD).

A dual beam FIB extends a FIB instrument's capabilities in various aspects: Progress during FIB cutting can be monitored by simultaneous SEM imaging. Gas injection systems (GIS) can be used in combination with the electron beam to deposit protection layers in a non-destructive manner. A piezo-controlled micromanipulator allows the transfer of the well-establish ex-situ lift-out preparation technique into the FIB chamber: In-situ lift-out can be done with higher reliability and has the advantage that the lamella can be further thinned after extraction from the sample surface.

A recent development in the field of FIB techniques is the use of a dual beam FIB for automated serial sectioning and imaging (3). For this technique first a SEM image is taken from a FIB cut cross-section. After that a slice of typically 20-50 nm thick material is removed from the cross-section by further FIB cutting. Then a SEM image is acquired again and the procedure is repeated several times. Such image series can be used as basis for 3D reconstruction and allow tomography on structures down to sub-50 nm feature size.

In the following sections we present applications of some of the above described techniques. The first example has been chosen as it covers all common TEM techniques for element analysis and illustrates that one has to proceed carefully during elemental analysis. The second example shows how the SiGe material specific electron energy loss spectrum can be used to obtain a compositional quantitative EFTEM/ESI mapping routine on a sub 10 nm scale that is calibrated by Rutherford back scattering (RBS). The third example features 3D analysis of defects in silicon devices by first imaging a 4 μm thick plan-view specimen at 400 kV for defect location and then by FIB cross-sectioning the plan-view specimen. The last example demonstrates the application of FIB tomography for characterizing partial adhesion / delamination at a chip / molding compound interface.

The analytical TEM investigations were performed in a Philips CM200 FEG applying 1nm STEM-EDX/EELS and electron spectroscopic imaging with a Noran HPGe EDX detector and an attached GIF. Defects in 4 μm thick silicon samples were located at 400 kV in a JEOL 4000EX. FIB preparation was performed in a Micrion 2500 single beam FIB or a STRATA 400 dual beam FIB, the latter one being equipped with an Omniprobe 200 in-situ manipulator, a gas injection system for carbon and platinum deposition and the FEI Auto Slice &View module for automated serial sectioning. Conventional TEM preparation was done by mechanical grinding and subsequent Argon milling in a Gatan PIPS.

EDX/EELS Investigation of the Worm Corrosion of Aluminum Interconnects

The processing of Al layers during the fabrication of Al interconnects for integrated silicon semiconductor devices includes patterning and cleaning steps. Both the reactive gases for patterning and the chemical solvents for cleaning are generally chlorine-based. Residues of chlorine may locally cause corrosion of the Al interconnects during subsequently performed water-based rinsing and cleaning steps. Corrosion decreases the interconnect diameter and is of course highly undesirable. The chlorine contaminated regions are typically decorated by worm-like outgrowths (Figure 1) which gave the phenomenon the name worm corrosion.

Since such "worms" have sizes below 1 µm and must not be exposed to water in order to avoid further corrosion, TEM preparation by FIB lift-out is the method of choice. Figure 2 shows a cross-sectional view of a worm pouring through a hole in a silicon oxide mask extending from the aluminum interconnect underneath. 1nm EDX measurements reveal that the worm consists out of aluminum oxide and contains a significant amount of chlorine (Figure 2).

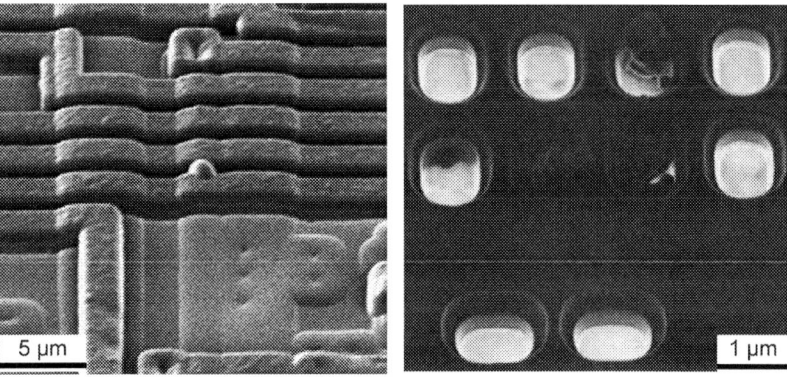

Figure 1. Corrosion worms at the edge of an Al interconnect (left) and pouring through a hole in a Si oxide mask from the underlying Al interconnect (right).

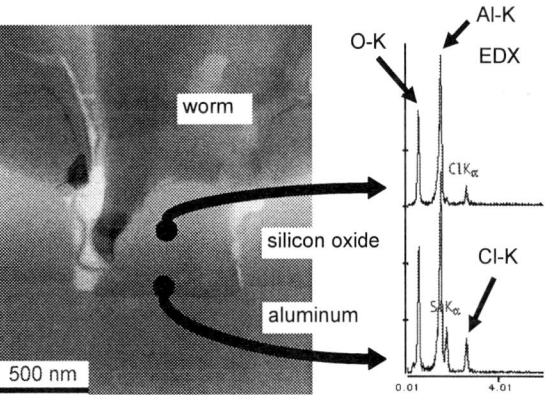

Figure 2. STEM BF image (left). EDX spectra of the worm formed by corrosion (right).

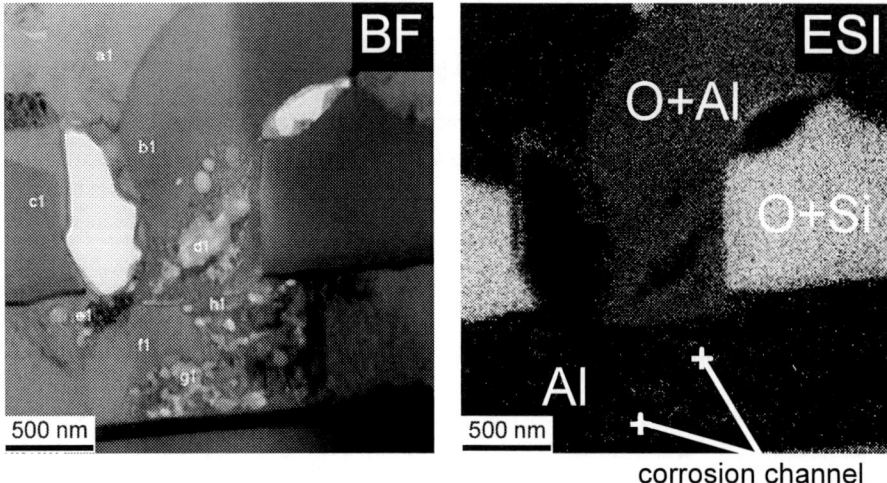

Figure 3. TEM BF image (left). RGB montage of electron spectroscopic edge jump-ratio images (O-K red, Si-K green, Al-K blue; right).

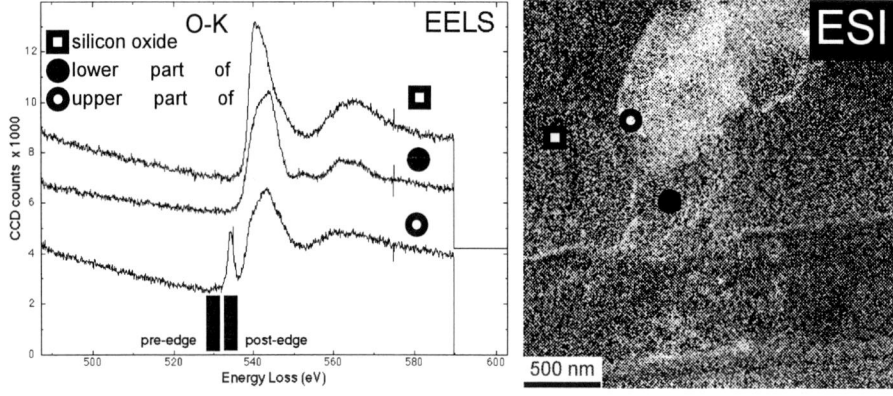

Figure 4. EELS spectra from the upper and lower part of a worm (left). Edge jump-ratio image with the O-K pre-edge feature depicting bright contrast only in the upper part of the worm (right).

As the lift-out TEM lamellae are thin enough for EELS analysis over areas, 5-10μm wide, EFTEM/ESI is a convenient and fast method to obtain elemental distribution maps for the elements oxygen, aluminum and silicon. Figure 3 shows a RGB montage of superimposed electron spectroscopic edge jump-ratio images which reveals the presence of aluminum oxide both within the worm and inside corrosion channels within the aluminum interconnect underneath. This proofs that the aluminum within the worm comes from the aluminum interconnect.

An interesting observation was made when reinvestigating a worm with an EDX proven high content of chlorine in its upper part after two weeks storage on air: The chlorine had vanished from the worm. In order to understand this observation the same TEM sample was submitted to EELS analysis. Figure 4 compares EELS spectra from the upper and lower part of the worm and from the silicon oxide mask. For the lower part of the worm and the silicon oxide the O-K edge depicts the typical shape for aluminum oxide and silicon oxide, respectively. For the upper part of the worm the O-K edge is aluminum oxide-like and additionally depicts a pre-edge feature. This pre-edge feature is so pronounced that it is suitable for ratio mapping: The ratio image reveals that the pre-edge feature is present throughout the upper part of the worm, though to different degrees (Figure 4).

O-K pre-edge features have been observed for several ceramic materials, and it has been reported that its intensity increases with the concentration of vacancies on the oxygen sub lattice (4,5,6). For corrosion worms the pre-edge feature is characteristic for regions with high chlorine content and appears after the disappearance of chlorine only. A possible explanation for this observation is that corrosion continues when the TEM lamella is not stored in vacuum by reaction with moisture from the surrounding air. This process modifies the aluminum oxide that was initially formed by water rinsing during device manufacturing. Chlorine disappears by out diffusion from the thin TEM lamella and redistributes across its surface. Thus it is not detectable anymore by EDX.

Figure 5. TEM BF image and EDX analysis (left). RGB montage of electron spectroscopic edge jump-ratio images (F-K red, O-K green, Al-K blue; right) .

One further interesting observation was made when investigating devices from later stages of device processing: Instead of aluminum oxide the worms consisted of aluminum fluoride (Figure 5). However, carefully performed EDX analysis revealed chlorine inside corrosion channels within the aluminum interconnect. The explanation for these observations is that chlorine-induced corrosion worms formed during early stages of device processing are exposed to fluor-rich solvents later on. In that way the chlorine-rich aluminum oxide phase is transformed into a more stable aluminum fluoride phase.

Germanium Concentration in Silicon-Germanium Layers

SiGe layers are used in state-of-the-art high frequency bipolar transistors in the base region and in the last few years have found application for carrier mobility enhancement in MOSFET channels. Moreover SiGe layers are used to form strained silicon which in turn increases mobility for sub 100nm MOSFETs. As the Ge content of the $Si_{1-x}Ge_x$ layer determines the physical parameters of the device there is a need to determine the Ge content x for layers with few nanometers in thickness.

A few years ago Pantel and coworkers (7) realized and reported a difference in the background signal of electron energy loss spectra for Si, Ge, and $Si_{1-x}Ge_x$ alloys. The difference in background signal is apparent between 50 and 100 eV, just in front of the $Si-L_{2,3}$ edge. The reason for this was assumed to be the influence of the delayed $Ge-M_{4,5}$

edge at 29 eV which is broad and extends into this energy range. ESI ratio images 90 eV/60 eV obtained from this region were shown to be insensitive to TEM specimen thickness variations, diffraction contrast effects, and poly SiGe grain orientation. The intensity could be calibrated with quantitative EDX measurements and was proposed for quantitative analysis.

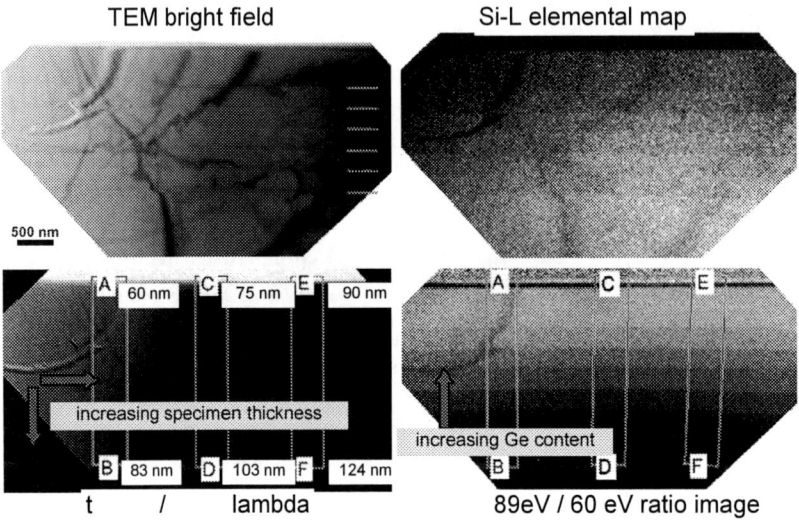

Figure 6. Energy-filtered imaging of SiGe step structure.

Here we revisit this method and give an estimation about the accuracy of the technique. For calibration a sample from a Si wafer was used that had deposited six SiGe layers with stepwise increasing Ge content. A Z contrast STEM image displays these layers in Figure 7. The Ge concentrations indicated in the image were obtained by RBS measurement. Figure 6 shows a typical bright-field image and a Si elemental map (ESI with Si-$L_{2,3}$ edge) of the layer structure. Diffraction contrast is visible in both images and the layers do not depict a strong contrast difference. Edge ratio images only show a subtle improvement. To study the dependence on specimen thickness and obtain a calibration curve, four different specimen regions were used for electron spectroscopic imaging. First t/λ images (t: specimen thickness, λ: mean free path of plasmon scattering) were recorded and then images at 89 eV and 60 eV (slit width 15 eV and largest objective aperture, i.e. about 20 mrad collection semi-angle). Under these conditions λ corresponds to approximately 150nm (8). Figure 6 shows one of the t/λ images. Specimen thicknesses at the marked positions were calculated and are inserted as labels in Figure 6. Three line scans were made across the SiGe layers for each of the t/λ images and the corresponding 89eV/60eV ratio images (Figure 6).

The results for line scans from point C to D are shown in Figure 7: The signal of the 89eV/60eV ratio image reproduces clearly the stepwise increase in the Ge content while the thickness changes strongly. Evaluation of all line scans of each of four specimen locations yielded that the signals of the 89eV/60eV ratio images fell almost exactly on top of each other whereas the thicknesses values varied between $t/\lambda = 0.2 - 0.8$, i.e., between 75 nm and 120 nm. Figure 8 displays the calibration curve for Ge concentrations between 0 and 30 at%. The upper and lower margin curves result from twelve different line scans in 89eV/60eV ratio images. Hence the Ge concentration can be determined with an accuracy of about ± 4.5 at% in the concentration range from 5 to 20 at % Ge.

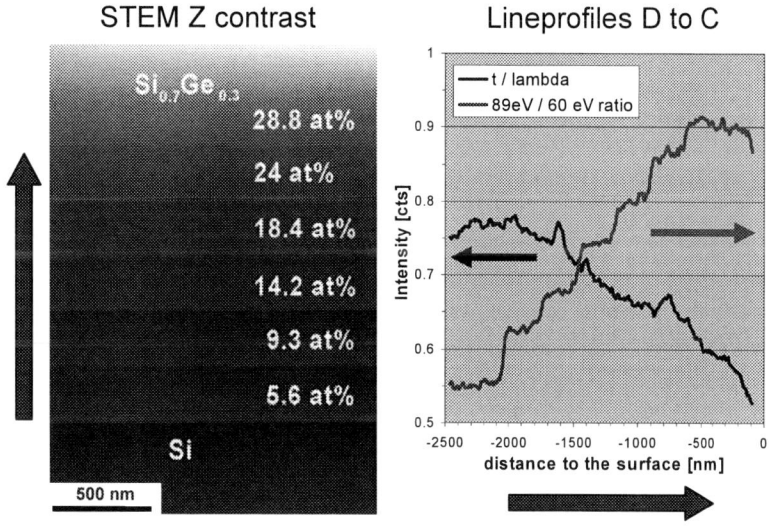

Figure 7. STEM and energy-filtered imaging of SiGe step structure.

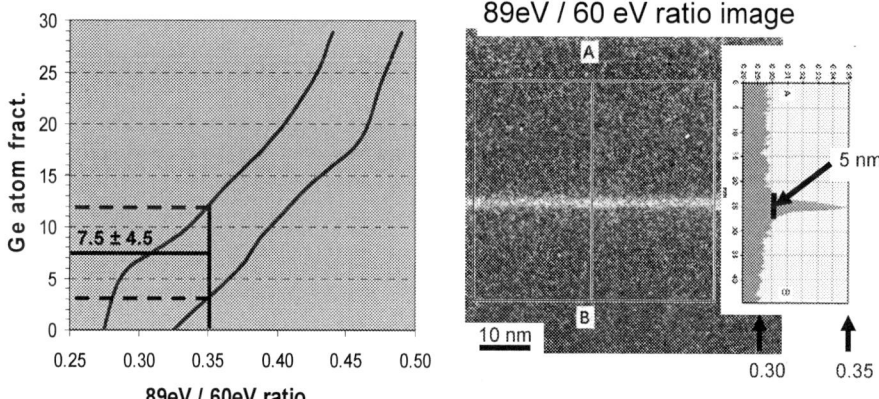

Figure 8. Calibration curve and experimental determination of Ge content.

The example of a Ge delta doping layer in silicon grown by chemical vapour deposition demonstrates the accuracy of the method. RBS yields that the equivalent to 0.9 mono-layer pure Ge is deposited. Figure 8 shows the 89eV/60eV ratio image. With help of the calibration curve the peak concentration is determined to be 7.5 ± 4.5 at%. The concentration at half maximum of the peak amounts to 3.5 at %. The contrast of the Ge delta spike is approximately 5 nm wide at the base. This delta peak can be considered as being smeared out over 20 (002) planes. By calculation this gives a surface concentration of about 0.9 mono-layers in consistence with the RBS results on the sample.

3D localization and analysis of defects in semiconductor devices

Electrical failures in semiconductor devices are routinely identified during testing. The specific electrical failure location can often be found by techniques delivering hot spots such as liquid crystal imaging, emission microscopy and others. TEM characterization of such failure sites often faces the problem that, e.g., a hot spot may have a lateral extension up to 10 µm whereas the electrically active defect, e.g., a dislocation in the near surface region of the silicon substrate, has a much smaller extension. Conventional TEM preparation of thin cross sections or plan-view specimens may therefore miss the actual defect. In the following we describe a preparation procedure that permits in many cases to localize the defects safely with higher resolution in a thick plan-view TEM specimen and then allows to prepare a TEM cross-section at the exact defect position from the plan-view sample.

The upper part of Figure 9 shows the adaptation of the conventional FIB preparation technique (also called H-Bar technique) to plan-view preparation of a typically 10 µm × 10 µm large area containing the, e.g. top most, 4 µm of a silicon device. The localization of the defects in the plan-view orientation is done by conventional TEM imaging at 400 kV which allows 4 µm thick silicon samples to be transmitted with sufficient contrast. Figure 10 shows a TEM bright field image of a defect that permits to confine the defect to an area of about 200 nm in diameter.

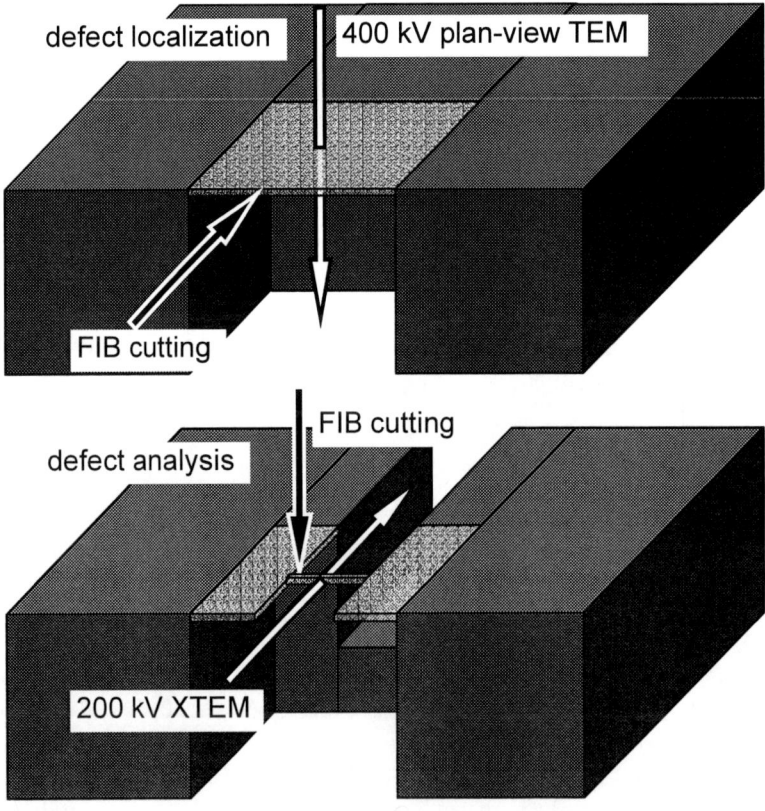

Figure 9. Subsequent steps of defect localization and analysis.

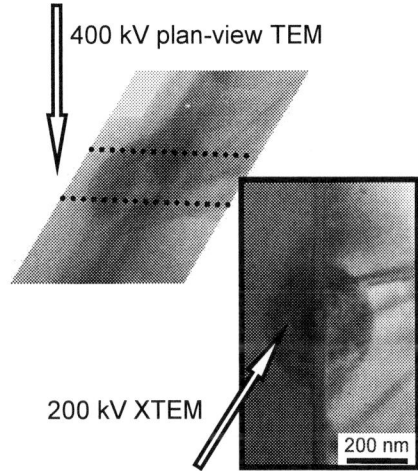

Figure 10. TEM imaging of the same defect in plan-view and cross-sectional orientation.

In order to proceed with cross-sectional preparation we rely on a 3 mm Cu half disk that has been used for wedge preparation of III/V semiconductors for a long time (9). Figure 11 shows how the half disk is prepared. For the procedure the silicon chip is attached to the flap of the grid prior to FIB plan-view preparation. The advantage is that during plan-view preparation and imaging the flap is parallel to the grid (flap in). Once the defect has been localized by 400 kV TEM the flap is put down (flap out) and one proceeds with cross-sectional FIB preparation and subsequent imaging (Figure 9). The second micrograph in Figure 10 shows the same defect in cross-section at a sample thickness of 500 nm. For other cases such cross-sections were successfully further thinned below 100 nm specimen thickness and the defect was submitted to elemental analysis in the TEM.

The presented technique has potential in the field of site-specific preparation for electron tomography or atom probe: The remaining part of the specimen containing the defect has the form of a bar. After extraction by in-situ lift-out it can be shaped to the form of a cylinder or needle by further FIB milling parallel to the bar axis.

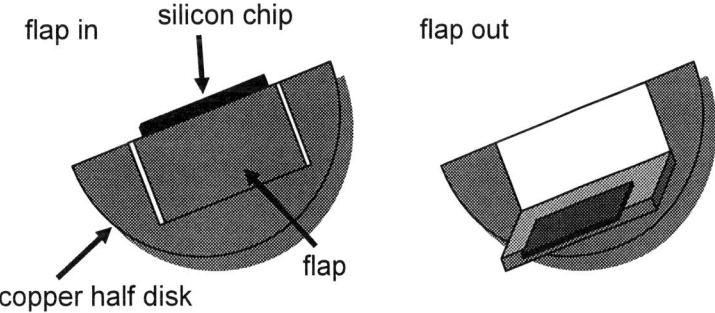

Figure 11. Specimen grid used for support of the silicon wafer during the two subsequent steps of the FIB preparation and TEM analysis.

3D reconstruction of delamination gaps by FIB tomography

Failure of semiconductor devices is often caused by delamination at interfaces within the packaged die. Examples are the detachment of a bond wire from the chip surface due to mechanical load or delamination at the chip / molding compound interface during thermal stressing, due to different thermal expansion coefficients. In many cases even prior to the application of any load the critical interface reveals partial delamination in form of voids. During thermal or mechanical stressing prior to complete delamination different degrees of delamination may occur including roughing of the interface and plastically deformation of the adjacent materials. Thus there is a need to characterize the delamination process on a sub μm-scale in a quantitative manner.

A common technique for the detection of delamination is scanning acoustic microscopy (SAM). Though SAM may detect gaps which are less than 50 nm wide its lateral spatial resolution is worse than 10 μm when penetration depths of several 10 μm are required. Thus automated serial sectioning and imaging with a dual beam FIB is a valuable option to perform a 3D characterization with a resolution below 50 nm.

For samples with partial delaminating or poorly adhesive interfaces mechanical stress during preparation should be avoided as far as possible. Therefore for the investigation of delamination between top-most chip layer and molding compound the device is ground from the chip side down to a final silicon substrate thickness of 20 μm prior to FIB/SEM analysis. This procedure is intended to minimize mechanical load of the interface as the interface is protected by the remaining silicon during the complete grinding procedure. For FIB serial sectioning and imaging one proceeds by FIB cutting cross-sections through the remaining silicon and the interface layers into the mold compound.

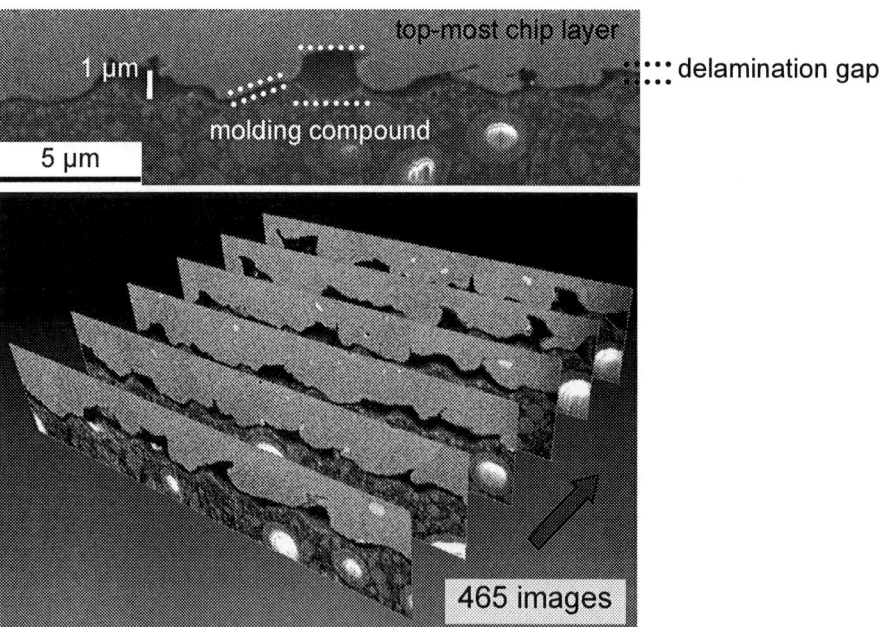

Figure 12. SEM imaging on FIB cut of delaminating interface and FIB serial sectioning and imaging.

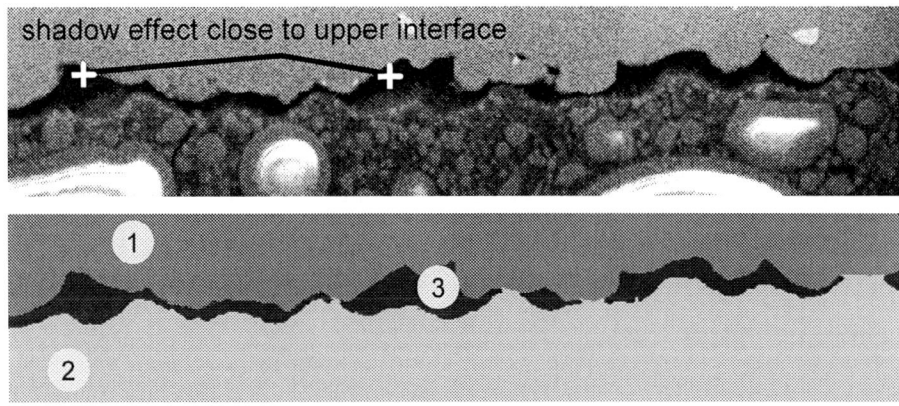

Figure 13. Shading effect during SEM imaging and sequence of reconstructed objects.

The upper part of Figure 12 shows an SEM image of such a FIB prepared cross-section. It is well discernible that both interfaces are very rough and that the width of the delamination gap strongly varies laterally. The lower part of Figure 12 shows 6 subsequent images from an sequence of 465 image acquired by automated serial sectioning and imaging in the dual beam FIB. The lateral pixel width is about 30 nm, the spatial separation of subsequent images, i.e. the slice width, amounts to about 60 nm.

3D reconstruction of this image series was performed using Mecury Computer Systems Amira ResolveRT Software following the usual procedure of image alignment and grey level segmentation (3). The oblique impingement of the electron beam onto the FIB cross-section, the angle between electron and ion column is 52° for the FEI STRATA 400, results in a shadow effect at the upper interface (Figure 13). Thus the direct reconstruction of the gap by segmentation was not possible. However, after separate segmentation and reconstruction of the adjacent layers the gap surface could be deduced (Figure 13).

Figure 14 shows an oblique view of the reconstructed gap surface with a lateral dimension of about 30 μm × 30 μm. In this view the gap interface on the chip layer side is oriented upward. Thus one can well recognize that the top-most chip layer displays a wave-like surface roughness. Contact areas of top-most chip layer and molding compound which correspond to "holes in the gap" are observed within valleys of these waves. As all spatial dimensions of the gap surface are calibrated it is suitable for geometrical measurements in 3D: In this sample the maximum gap width is about 2.5 μm and delamination rate is 84 %, i.e., only 16 % of top-most chip layer and molding compound interface have contact!

Conclusion

The combination of focused ion beam and high-end transmission electron microscopy techniques for sample preparation, and imaging and elemental analysis opens new perspectives and approaches to materials analysis problems.

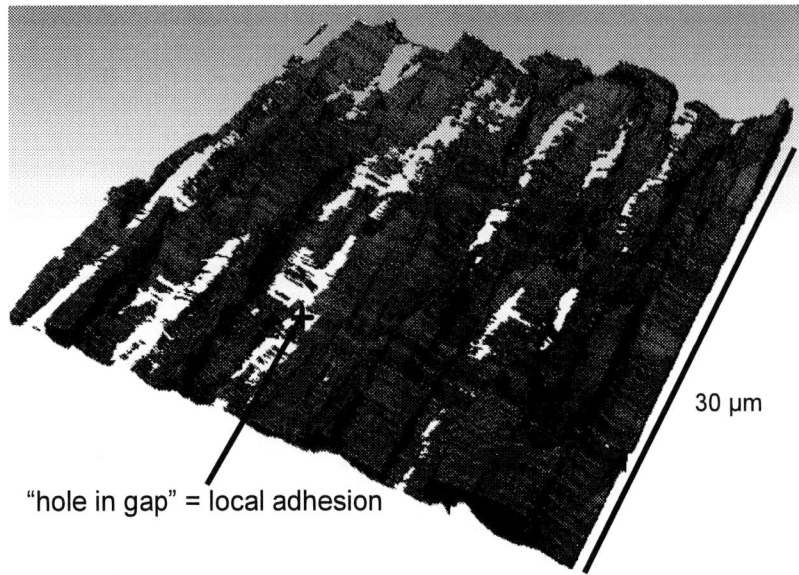

Figure 14. View on the reconstructed gap surface (the interface towards the top-most chip layer is oriented upward).

Acknowledgments

The authors wish to thank V. Klüppel and T. Ohnemus for long term pleasant and inspiring collaboration, F. Scrofani for support with the 3D reconstruction software and M. Leicht from Infineon Technologies for the fruitful collaboration.

References

1. M. Tence, M. Quartuccio, C. Colliex and Smith David J., *Ultramicroscopy*, **58** (1), 42 (1995).
2. J. Mayer, L. A. Giannuzzi, T. Kamino and J. Michael, in *Focused Ion Beam Microscopy and Micromachining / May 2007*, C. A. Volkert and A. M. Minor, Editors, VN 32-5, p. 400, MRS Bulletin, (2007).
3. M. D. Uchic, L. Holzer, B. J. Inkson, E. L. Principe and P. Munroe, in *Focused Ion Beam Microscopy and Micromachining / May 2007*, C. A. Volkert and A. M. Minor, Editors, VN 32-5, p. 408, MRS Bulletin, (2007).
4. Zhang H. and Dravid V. P., *J. Am. Ceram. Soc.*, **76** (5), 1143 (1993).
5. P. A. van Aken, B. Liebscher and V. J. Styrsa, *Phys Chem Minerals*, **25** (7), 494 (1998).
6. Th. Höche, H.-J. Kleebe, F. Schrempel, and W. Wesch, *Phil. Mag. Lett.*, **82** (11), 599 (2002).
7. R. Pantel, S. Jullian, D. Delille, D. Dutartre, A. Chantre, O. Kermarrec, Y. Campidelli and L.F.T.Z. Kwakman, *Micron*, **34** , 239 (2003).
8. D. R. G. Mitchell, *Journal of Microscopy*, **224**, 187 (2006).
9. H. Cerva, H. Oppolzer, *Prog. Crystal Growth and Charact.*, **20**, 231 (1990).

Characterization of Nickel-related Defects in Thin SOI Substrates after Thermal Treatment

I. Rink, C. Emons

NXP-Semiconductors, Nijmegen 6534 XE, Gerstweg 2, Netherlands

By means of thermal treatment of blanket wafers, the impact of nickel contamination (applied on the wafer surface) on defect formation in Silicon on Isolator (SOI)-material during oxidation has been studied. Methods like determination of defect density by microscopic control after extended vapor phase decomposition and preferential etching, minority carrier lifetime measurements and vapor phase decomposition - Inductively Coupled Plasma Mass Spectrometry were used to characterize the effect and the redistribution of nickel through the handle wafer after oxidation.

It is shown that with the same nickel concentration at the wafer surface the highest defect density is caused in the thinnest SOI-layer after oxidation not dependent from the thermal budget. Furthermore, it is found that nickel diffuses through the buried oxide which might relax the allowable limits of nickel at the wafer frontside before oxidation, because it is after oxidation dissolved in the handle wafer. However, the backside contamination with nickel becomes more critical because by the reverse diffusion nickel can reach the frontside and form defects.

Introduction

In the past nickel (Ni) has proven to be a very critical metal for SOI substrates, causing defects when present on the surface at very low levels before oxidation [1, 2]. Levels larger than 4E9 at/cm^2 already result in defects and yield problems in SOI - IC processes using SOI-films (top-Si) of 1.5μm and buried oxide (BOX) of 1μm. Latest generations SOI-IC processes use thinner top-Si and thinner BOX layers in lower thermal budget processes. At the same time critical metal levels are decreasing according to the ITRS-roadmap [3]. In this context the following questions have to be asked:

1. How sensitive is a thinner top-Si layer for NiSi$_2$- formation due to Ni-contamination on the wafer surface before oxidation?
2. What role plays the BOX-layer thickness and does it still function as diffusion barrier?
3. Will backside contamination cause defects in the top-Si regarding thinner BOX-layer?
4. Which impact has the thermal budget in this matter?

Experimental details

The focus of this study is the application of representative tests, so-called physical short-loops on blanket wafers to simulate the impact of nickel contamination. A large variety of common SOI-material with variations in thickness of the top-Si and BOX-layer in comparison with FZ-bulk material is used. In TABLE I all types of wafers under study including their properties are summarized.

TABLE 1: Overview of wafer material

variant	Type of handle wafer	Thickness top-Si	Thickness BOX	Type of top-Si
FZ	125mm, FZ , <100>, p-type, 150Ωcm	n.a.	n.a.	n.a.
FZ	150mm, FZ , <100>, p-type, 150Ωcm	n.a.	n.a.	n.a.
1500/3000	125mm, CZ, <100>, n-type, 2Ωcm	1500 nm	3000 nm	CZ, <100>, p-type, 18Ωcm
1500/1000	125mm, CZ, <100>, n-type, 2Ωcm	1500 nm	1000 nm	FZ, <100>, p-type, 18Ωcm
180/200	150mm, FZ, <100>, n-type, 200Ωcm	180 nm	200 nm	FZ, <100>, p-type
680/200	150mm, FZ, <100>, n-type, 200Ωcm	180 +500 nm	200 nm	FZ, <100>, p-type + EPI, n-type, 0.85Ωcm
200/200	150mm, CZ,<100>, p-type, 18 Ωcm	200 nm	200 nm	FZ, <100>, p-type
200/400	150mm, CZ,<100>, p-type, 18 Ωcm	200nm	400 nm	FZ, <100>, p-type,18 Ωcm

The physical short-loops follow in principle a typical process flow:

- Cleaning of the wafers to create a chemical oxide at the surface or to create a barrier oxide by thermal oxidation, respectively
- Intentional contamination with Ni at several concentrations, at the frontside (FS) or at the backside (BS) of the wafer, using the spin-dry method
- Oxidation of the wafers under several conditions
- Characterization of the impact of the treatment using Minority Carrier Lifetime determination (MCLT), Vapor Phase Decomposition-Inductively Coupled Plasma Mass Spectrometry (VPD-ICPMS), Forced-VPD (FVPD)(4) and preferential etching (PE) with Wright etchant (5) in combination with optical microscopy in Differential Interference Contrast mode (DIC)

In order to cover different impact situations four experiments were set up. In experiment 1 the sensitivity of thinner top-Si (variants 180/200 and 680/200) in comparison to thick top–Si (variants 1500/3000 and 1500/1000) was studied. With experiment 2 the possibility of oxide as diffusion barrier for Ni is checked (variant 680/200). In experiment 3 this issue is continued by contaminating the backside (BS) of the wafers with Ni (variants 200/400 and 680/200) and check the effect at the frontside (FS). Experiment 4 evaluates the impact of a lower thermal budget in combination with thin top-Si (variants 180/200, 200/200 and 680/200). Further details are summarized in TABLE II.

TABLE II: overview of the process conditions

experiment	Barrier oxide	Ni contamination/place	Oxidation
1	no	4 E10 at/cm^2 / frontside	1100°C, 1h, dry, 125 nm
2	1100°C, 1h, dry, 125nm	4 E9 at/cm^2 / frontside	1100°C, 1h, dry, 125 nm
3	no	3 E12 at/cm^2 / backside	1100°C, 1h, dry, 125 nm
4	no	1.4 E10 at/cm^2 /frontside	950°C, 30min, dry, 20 nm

In all experiments clean wafers also were processed according to the short-loop flow (excluding Ni-contamination) to have a baseline of each process.

Results and discussion

According to reference (1) SOI-material with thinner top-Si is expected to be more sensitive for certain nickel contamination levels to form $NiSi_2$-precipitates during thermal treatment due to the concentration effect in the top-Si. However, a simultaneous reduction of BOX-layer thickness and the reduction of the thermal budget might relieve the impact.

Sensitivity of top-Si thickness (Experiment 1)

Based on our first order model (1, 2) to predict the critical surface level for several elements necessary to cause defects under various oxidation conditions the critical Ni-

Figure 1: micro-photographs of "Ni-defects" after FVPD (left), after PE (centre), close-up after FVPD (right)

concentrations before oxidation to form precipitates during the chosen oxidation (dry, 1100°C, 1h) were estimated for the thinner SOI-variants in this experiment at 3.5E8 at/cm^2 for variant 180/200 and at 1.7E9 at/cm^2 for variant 680/200. Because of the chosen Ni-concentration of 4E10 at/cm^2 all SOI-variants should clearly show Ni-related defects after oxidation, which is indeed the case. For the variants 1500/3000 and 1500/1000 the defects are only visible after PE, but not after FVPD. FVPD (4) is a method to dissolve $NiSi_2$-defects and make them simultaneously visible. Both of the variants with thinner top-Si show after FVPD clearly microscopic visible defects. After PE only defects are observed for variant 680/200 at a considerably higher density then after FVPD. For variant 180/200 it was not possible to detect defects after PE, because the top-Si layer was completely destroyed, possibly due to the high concentration of defects. In Figure 1 examples of "Ni-defects" after FVPD (variant 180/200), after PE (variant 680/200) and a close-up of such a defect after FVPD for variant 680/200 are presented.

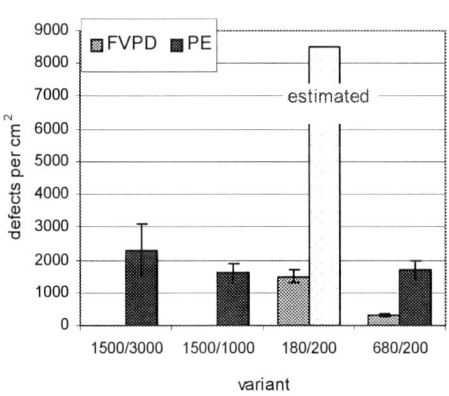

Figure 2: DD after FVPD and PE per cm^2

To quantify the results, the defect density (DD) at several sites on the wafer is determined and the average values of DD per cm^2 are presented in Figure 2. In a previous study (6) we tried to compare defects after FVPD and PE, and found a much higher defect density after PE than after FVPD. This can be explained by the Ni precipitation mechanism, which starts at the BOX/SOI − interface. Not all defects are reaching the surface implying that they are not all dissolved by the FVPD-procedure.

This effect explains that for the variants with the thick top-Si no defects are found after FVPD. Assuming the same relation for variant 180/200 between DD after FVPD and after PE as found for variant 680/200 we can estimate the defect density after PE for the 180/200 variant depicted in Figure 2 as dotted column.

Another effect is shown by the VPD/FVPD-ICPMS results (see Figure 3). The resulting nickel-concentrations after oxidation are much lower than before oxidation at the surface. The percentage of the recovered Ni is variant dependent. For FZ-material with very low recovery of nickel it can be assumed that a huge part of the Ni is distributed in the bulk of the wafer, since it is a rather fast diffusing metal. Only a small part is situated in the grown oxide and at the interface oxide/silicon. For the SOI-variants 1500/3000 and 1500/1000 it is known that the BOX-layer functions as a diffusion barrier under the chosen

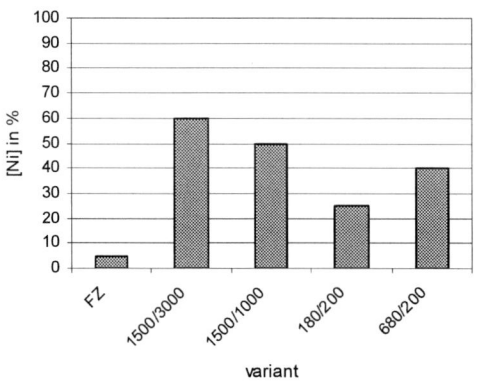

Figure 3: [Ni] in % due to VPD/FVPD-results after oxidation

oxidation conditions (1). Therefore, more than 50% of the initial Ni is recovered. The other part forms defects at the interface BOX/top-Si. A part of them is obviously too small to reach the surface of the top-Si and can, therefore, not be collected after VPD. This is different for the SOI-variants with thinner top-Si regarding the dimension of the defects. However, the recovery of Ni is in this case even lower (between 25 and 40%) than for the variants with thick top-Si. This effect points to the option, that the BOX-layer is not thick enough to function as effective diffusion barrier. That means the Ni-concentration available in the top-Si to form "Ni-defects" is reduced with respect to the initial concentration. In summary the results prove, that SOI-material with thinner top-Si is more sensitive than that with thick top-Si.

Impact of oxide as diffusion barrier for Ni during oxidation (Experiment 2)

For this experiment only SOI-variant 680/200 was used. Some of the wafers were

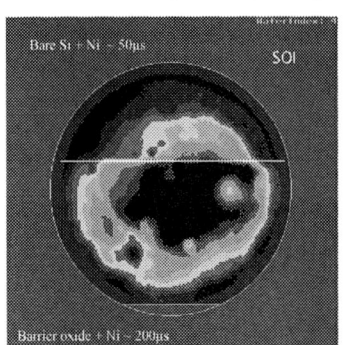

Figure 4: MCLT-image of a SOI-wafer after 2nd oxidation (barrier oxide partly removed before Ni-contamination)

additionally oxidized (barrier oxide) before Ni-contamination and second oxidation. The initial Ni-concentration of 5E9 at/cm^2 is in the range of the critical concentration to form "Ni-defects" in SOI-material with thick top-Si (1.5µm). Crystal defects were found on all Ni-contaminated SOI-wafers, however, not homogeneously distributed and in most cases these are very tiny. Therefore, it is very difficult to quantify a defect density.

Since these defects are also found on wafers with barrier oxide, we can conclude that the chosen oxide was not able to function as an effective diffusion barrier for Ni.

That the barrier oxide hampers metal diffusion in general is shown by MCLT. After the 2nd oxidation, the MCLT-levels decreased from about 400µs to 200µs for wafers with barrier oxide and to about 50µs for wafers without barrier oxide. In the MCLT-mapping of a SOI-wafer, where the barrier oxide partly is removed before Ni-contamination and the 2nd oxidation (Figure 4), this impact is very well visible. This effect is evidently caused by iron, according to the VPD-ICPMS – results.

<u>Ni-contamination of wafer backside before oxidation (Experiment 3)</u>

By contaminating the wafer backside, it is possible to look which role the BOX-layer thickness as well as the top-Si layer thickness will play in relation to "Ni-defect" formation under known thermal treatment conditions. Intentionally contaminating the backside without contaminating the front-side was the challenge of this case. It was only possibly by using a so-called low-contact chuck at the spinner. The Ni-concentration at the front-side determined by means of VPD-ICP-MS and FVPD-ICP-MS after oxidation shows a clear relation to the presence of Ni at the backside of the wafers and to each top-Si-variant variant as depicted in Figure 5. Ni-concentrations at the frontside of not contaminated wafers, but also of Ni-contaminated FZ-wafers are neglectible. That means, in contaminated FZ-wafers the Ni is distributed over the whole bulk of the wafer and Ni-

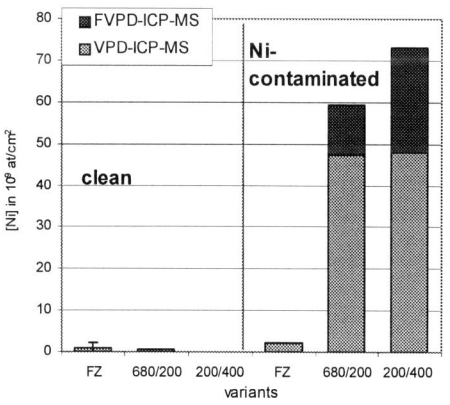

Figure 5: [Ni] determined on FS of the wafers by VPD/FVP-ICPMS

contaminated SOI-wafers show concentrations in the range of 5 to 8E10 at/cm^2, which is above the critical level to form defects. The differences between the SOI-variants after FVPD point to the presence of more defects on variant 200/400. This is indeed the case, as can be seen in Figure 6. At variant 200/400 not only more defects are observed, but also varying in size, whereas at variant 680/200 only large defects are visible.

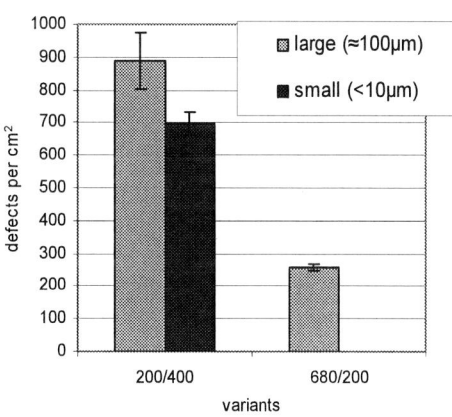

Figure 6: DD after FVPD per cm^2

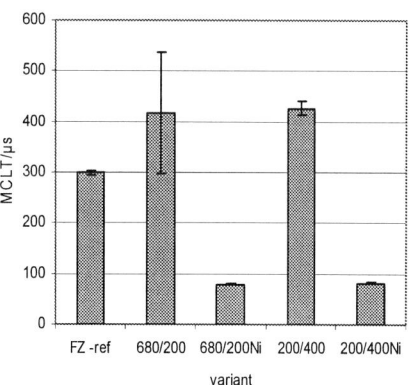

Figure 7: MCLT- average levels per variant

It also emphasizes that Ni under the chosen conditions does not evaporate during oxidation and cross-contaminate other wafers in the furnace.

Besides, the minority carrier lifetime of the SOI-variants is determined by the Ni-level in the handle wafer, which is in case of Ni-contamination at the BS clearly decreased. The MCLT-level comparison of Ni-contaminated and clean wafers is presented in Figure 7.

In summary, the results of this experiment reveal that BOX-layers of < 400nm do not form an effective diffusion barrier for nickel. Under these conditions the thickness of the top-Si mainly determines the number of "Ni-defects". It further implies, that SOI-material with thin BOX/top-Si is also vulnerable to Ni-contamination at the BS before thermal treatment.

Impact reduction of thermal budget (Experiment 4)

According to our model (1,2) using 1.4E10 at/cm^2 nickel as concentration for intentional contamination before 950°C oxidation should lead to "Ni-defects" (1E9 at/cm^2 to 4E9 at/cm^2 estimated as critical concentrations). Indeed at all SOI-variants the usual "Ni-defects" are observed however with smaller dimensions than in experiment 1. Determination of DD after FVPD shows a clear relation with the SOI-variant as demonstrated in Figure 8. Also in this case the variant with the thinnest top-Si (180/200) has the highest DD-level. It demonstrates, how sensitive thin top-Si is even in combination with low thermal budget. The trend found for the DD is confirmed by the results of VPD/FVPD-ICPMS given in Figure 9. The recovered Ni-amount after VPD and FVPD, mainly representing dissolved "Ni-defects", is considerably higher for the SOI-variants with thinner top-Si with 85 and 71%, respectively, than that for variant 680/200 with about 16%.

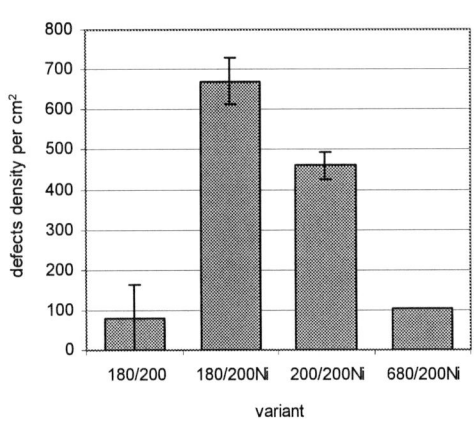
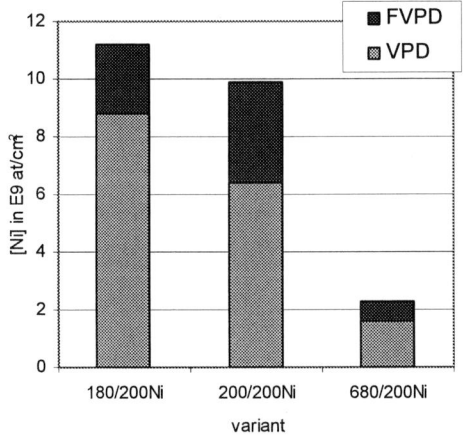

Figure 8: DD per cm^2 after 950°C oxidation and FVPD

Figure 9: [Ni] determined by VPD/FVPD-ICPMS after 950°C oxidation

Since all variants have the same BOX-layer thickness we can assume, that the conditions for Ni-diffusion through the BOX-layer are comparable for the three variants, explaining the concentration differences only by top-Si thicknesses.

Conclusion

By means of physical short-loops and physical & chemical measurements (determination of defect density, minority carrier lifetime, contamination analysis) it is possible to study the conditions, which lead to $NiSi_2$-precipitation on SOI during thermal treatment.

The study proves that SOI-material with a thinner top-Si/BOX-layer combination as typically used in the latest generation SOI-IC processes, is also sensitive to Ni-contamination on the frontside of the wafers before thermal treatment as SOI-material with a thicker top-Si/BOX-layer combination when oxidized under the same conditions.

By means of using a barrier oxide or Ni-contamination at the backside before oxidation it is shown that the thinner BOX-layer (< 400nm) does not function as effective diffusion barrier and is not able to avoid Ni-defect formation. This makes SOI-material with thinner top-Si/BOX-layer combination more vulnerable to Ni-contamination at the backside compared to that with thick top-Si/BOX-layer combination. Besides, this material is not covered with oxide at the backside like SOI-material with thick top-Si/BOX-layer combination.

Oxidation with a lower thermal budget typically used in latest generation SOI-processes shows a similar correlation between top-Si thickness and defect density. Defects can clearly be recognized, although they are smaller than after oxidation with higher thermal budget.

Acknowledgments

Thanks to L. Winters for performing the oxidations, E. Koelemij and H. Eberhard for the VPD-ICPMS analyses and to J. Bensussian from SOITEC for disposition of several types of SOI-wafers. This work is supported by Medea + - project 2T102 "HYMNE".

References

1. I. Rink et al. in *Crystalline defects and contamination: Their impact and control in device manufacturing III, DECON 2001* PV 29, p. 241 The Electrochemical Society Proceedings Series, Pennington, NJ (2001).
2. I. Rink et al. in *Ultra Clean Processing of Silicon Surfaces V*, UCPSS 2002 proceedings, p. 93
3. Semiconductor Industry Association (SIA), *International Technology Roadmap for Semiconductors 2006 Edition,* International SEMATECH Austin 2006
4. I. Rink, in *Ultra Clean Processing of Silicon Surfaces V*, UCPSS 2002 proceedings, UCPSS 2002, p.171
5. M. Wright-Jenkins et al., *J. Electrochem.Soc.* **124**(5), 757 (1977)
6. I. Rink, M. Chermin, *Philips internal report, RNR-R52-02/DJ0029,* 2002

ECS Transactions, 10 (1) 117-126 (2007)
10.1149/1.2773982 ©The Electrochemical Society

Detailed Photocurrent Analysis of Iron Contaminated Boron Doped Silicon by Comparison of Simulation and Measurement

M. Rommel[a], A. J. Bauer[a], and H. Ryssel[a,b]

[a] Fraunhofer Institute of Integrated Systems and Device Technology (IISB),
Schottkystrasse 10, 91058 Erlangen, Germany
[b] Chair of Electron Devices, University of Erlangen-Nürnberg,
Cauerstrasse 6, 91058 Erlangen, Germany

In this work, an extensive injection level dependent carrier lifetime study on intentionally iron contaminated boron-doped silicon has been performed by using the Elymat carrier lifetime method. The influence of both, iron and boron concentrations is investigated. Results from both Elymat measurement modes are considered and critically compared with simulations. The results clearly indicate that for low injection conditions, surface passivation with diluted HF is not sufficient whereas so-called electrostatic passivation allows for correct lifetime measurements. For the first time, results from both Elymat measurement modes are modeled consistently. The results prove that using optimized measurement and evaluation procedures the Elymat method is an appropriate technique for the quantitative determination of iron in case of iron being the relevant contaminant in boron doped silicon.

Introduction

In semiconductor industry, carrier lifetime measurements are routinely used for qualitative wafer and process characterization in order to detect possible bulk contamination with a high lateral resolution (e.g., 0.5 mm) on a full wafer scale. If contamination has been detected and its origin is not obvious (e.g., from the image of the lifetime map) other time consuming and elaborate techniques like deep level transient spectroscopy (DLTS) (1) have to be applied which are dedicated to the identification and quantification of these contaminants. For several important cases, however, carrier lifetime metrology also offers the possibility of quantitative determination of the trap concentration. Injection level dependent lifetime measurements (2) and the analysis of the temperature dependence of the carrier lifetimes (3) enable the identification and quantification of the dominant contaminant. Yet, for such analyses a very detailed understanding of the applied lifetime method as well as of the contaminant of interest is required. In this work, the Elymat carrier lifetime method (4) was used for an extensive injection level dependent carrier lifetime study on intentionally iron contaminated boron-doped silicon with different boron concentrations. Although results of Elymat measurement for iron contaminated silicon samples have been published before (5, 6, 7) a consistent modeling and interpretation of results from both Elymat measurement modes is presented here for the first time. Contrary to former work (7), no dissociation of iron-boron pairs is performed. Another focus of this study is the accurate determination of Elymat carrier lifetimes at low injection conditions. It will be shown that efficient surface passivation is the key in this injection regime.

117

Carrier Lifetime Measurement Method

Elymat Method

The Elymat technique (4) allows for the determination of photo induced diffusion currents I_{FPC} and I_{BPC} at both, the irradiated surface (so-called frontside photocurrent or FPC mode) and the backside of the sample (so-called backside photocurrent or BPC mode) (see figs. 1 and 2). During the measurement, the wafer is located in a double electrolytic cell. Diluted HF is used as electrolyte. On the wafer side where the current is not extracted, the electrolyte passivates the wafer surface. On the opposite wafer side, a reverse bias is applied using the electrolyte as a full wafer area contact. This results in a space charge region (SCR) which collects the photo induced minority charge carriers resulting in either I_{BPC} or I_{FPC}. The current level directly depends on the number of recombination centers in the sample. Therefore the currents are a direct measure for the carrier diffusion length L or lifetime τ. Routinely, the BPC mode is used for material quality or process monitoring because in FPC mode mainly the region near the surface influences the measured current. In BPC mode, however, the whole silicon bulk as well as defects at the irradiated surface affect the measured current. By changing the power of the scanning laser, the injection level dependence of the lifetime can be determined. In the Elymat system, the laser power is given by a so-called photo current I_{Photo} which equals the FPC current of a silicon sample with very high carrier lifetime at the given laser power.

Superimposed to the photo induced diffusion current is a dark current I_{dark} generated in the SCR without illumination. Depending on the defect density in the SCR which extends over the whole wafer surface, I_{dark} can be much higher than the diffusion current. To ensure optimum signal-to-noise conditions and to check for non ideal behavior of the sample an IV curve for both, diffusion and dark current is measured (fig. 6) before the actual lifetime mapping of the sample. From this IV curve, the working point in terms of V_{BPC} (or V_{FPC}, respectively) is defined.

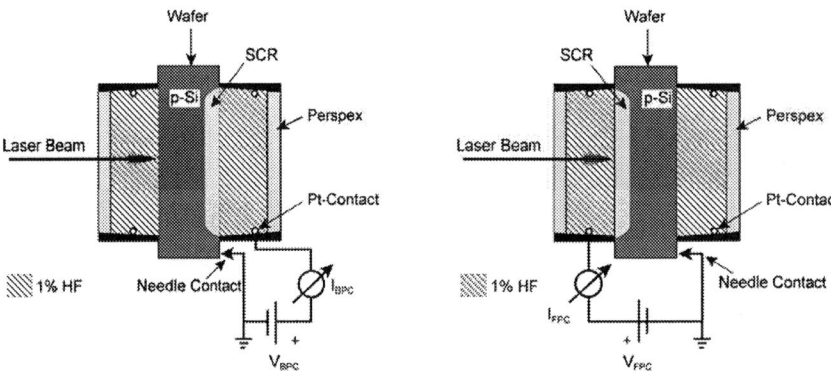

Figure 1. Elymat setup in BPC configuration.   Figure 2. Elymat setup in FPC configuration.

Elymos Method

The Elymos method (8, 9) is a modification of the Elymat method and its main advancement is the option for the additional characterization of the insulator/silicon interfaces in terms of interface state density and insulator charge. In the Elymos method, the photo induced diffusion current is measured in BPC mode. With the additional voltage

V_{EIS} applied across the electrolyte/insulator/semiconductor (EIS) structure it is possible to sweep the semiconductor surface potential from accumulation to inversion conditions (fig. 3). In accumulation the surface or interface recombination is eliminated very efficiently because all minority carriers are repelled from the surface (i.e., so-called electrostatic passivation). Therefore the bulk carrier lifetime can be determined very accurately under these conditions. Instead of diluted HF, acetic acid is used as electrolyte.

Data Evaluation and Simulation

The evaluation of the photo induced diffusion currents is based on a 1D diffusion problem with Elymat specific boundary conditions assuming a constant carrier diffusion length throughout the sample (4, 10). The exact solution of this problem requires iterative computing. The carrier lifetimes can only be calculated analytically if several simplified approximations are assumed as it is the case in the standard commercial Elymat evaluation software (11). The most important approximation for the FPC mode is that the sample thickness should exceed the diffusion length by far. The most important approximation for the BPC mode is that the penetration depth of the photons should be much smaller than the diffusion length.

The above stated assumptions, however, are not fulfilled in many practical cases (e.g., in FPC mode the assumed approximation is not fulfilled for carrier lifetimes above 50 µs, fig. 4). Therefore the exact iterative solving of the 1D diffusion equation is performed in this work in order to extract the Elymat carrier lifetimes from both, measured and simulated diffusion currents. The comparison of lifetimes from experiment and simulation is then used for the interpretation of the results.

For the simulations of the Elymat method, the commercial physical device simulator DESSIS (12) was applied. In addition to iron boron pairs (FeB), the small amount of interstitial iron (Fe$_i$) which is present in boron-doped silicon is also accounted in the modeling (13). The injection level dependence of the carrier recombination due to iron boron pairs with their relevant energy levels at 0.29 eV below the conduction band edge (acceptor level) and 0.10 eV above the valence band edge (donor level) is implemented using the model of Choo (14). Auger recombination is accounted as published by Kerr and Cuevas (15).

The values of the capture coefficients of the iron related energy levels strongly influence the absolute values of simulated carrier lifetimes as well as the lifetime dependence on injection level and boron concentration. Therefore it is important to note that in this work independently determined capture coefficients for the modeling of carrier recombination due to iron contamination were used. Compared to our recent work (13) where the capture coefficients were extracted from a detailed investigation on lifetime data (mainly measured using SPV and µ-PCD techniques) published for iron contaminated p- and n-type silicon additional published data sets were also examined (14, 15, 16, 17) which resulted in a minor change (maximum of 10%) of the capture coefficient values (see table I). The values presented here and in our recent work agree very well with the recently published data from an extensive QSSPC lifetime study by Rein (3).

TABLE I. Relevant iron related energy levels and their corresponding capture coefficients.

Iron Related Energy Level	Electron Capture Coefficient in cm³/s	Hole Capture Coefficient in cm³/s
Fe$_i$: E_V+0.38eV	$5.45 \cdot 10^{-7} \pm 4.5 \cdot 10^{-8}$	$1.07 \cdot 10^{-9} \pm 3.5 \cdot 10^{-10}$
FeB: E_V+0.10eV	$1.15 \cdot 10^{-6} \pm 5.0 \cdot 10^{-7}$	$1.23 \cdot 10^{-6} \pm 3.6 \cdot 10^{-7}$
FeB: E_C-0.29eV	$5.31 \cdot 10^{-8} \pm 2.4 \cdot 10^{-8}$	$2.25 \cdot 10^{-7} \pm 3.6 \cdot 10^{-8}$

2D simulation with rotational symmetry was chosen to include lateral diffusion whilst keeping the computational effort at a reasonable level. Using a numerical device simulator allows to properly account for the different injection levels and thus different local carrier lifetimes throughout the sample. This is due to the charge carrier diffusion and the carrier collection at the SCR. Consequently Elymat lifetimes determined as described above are effective lifetimes. Below a certain low injection level, however, carrier lifetimes are constant and the corresponding Elymat carrier lifetime can directly be compared with results from analytical models where constant injection is assumed (14, 20, 21) (see fig. 4). The very good agreement between the analytical and numerical results for low injection conditions for BPC and FPC mode (fig. 4) point out that both, the improved evaluation procedure (i.e., exact iterative solving of the 1D diffusion equation) and the geometrical and numerical setup of the simulator were properly selected.

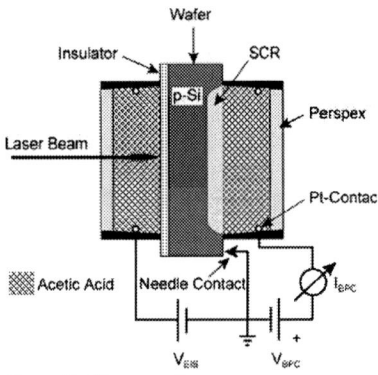

Figure 3. Elymos setup.

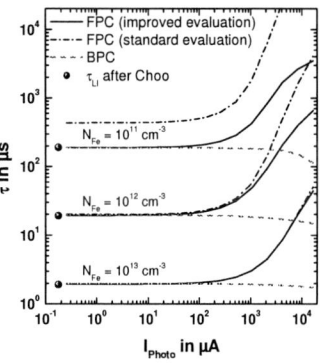

Figure 4. Lifetimes from simulated diffusion currents using standard (11) and improved evaluation for BPC and FPC mode. The boron concentration is $1.4 \cdot 10^{15}$ cm^{-3}. τ_{LI} represent calculated lifetimes using the model from Choo (14).

A detailed look on the results of the simulations can explain the different behavior of FPC and BPC lifetimes with increasing laser power I_{Photo}. In FPC mode, an internal electrical field is created due to ambipolar diffusion which results in a drift current in addition to the diffusion current. Therefore the total current increases but the evaluation procedure is based on the assumption of diffusion currents only and assumes increased diffusion which results in an increase of the evaluated apparent lifetime.

Sample Preparation

Two groups of samples (A and B) were prepared for the analyses starting with (100) MCZ grown boron doped silicon wafers of 150 mm diameter and a thickness of 675 µm. Group A consists of samples with different specific resistances and different iron contamination levels. All samples of this group were iron contaminated using the spin-on technique (22) and spiked solutions with different iron concentrations. After a drive-in, rapid thermal anneal at 1000° C for 120 s a diffusion anneal at 1100° C for 8 h in a furnace with N_2/H_2 atmosphere was applied. For all these samples, an inhomogeneous lateral iron distribution was observed which was probably due to the spin-on process itself. Therefore the absolute iron concentration for each sample is not known and can differ. Consequently, the different samples are distinguished and in the following referred to by their relative contamination level (medium, high, and very high concentrations, table II).

For the samples of group B, iron contamination was performed using ion implantation of different Fe^+ ion doses at 700 keV through a thermal silicon dioxide layer of 45 nm and subsequent rapid thermal annealing at 1100° C for 180 s. For the implantation an Al foil with a circular aperture with a radius of about 1.25 cm was used as a mask to enable the accurate determination of the iron limited carrier lifetime by comparison of the lifetimes outside and inside of the implanted region. After a forming gas anneal at 430° C to decrease the interface state density of the SiO_2/Si interface, the oxide on the wafer backside was removed by wet chemical etching. For samples of group B electrostatic passivation of the wafer surface can be used with defined and stable thermal SiO_2 layers. Due to the SiO_2 layers, however, FPC measurements can not be performed. Table II summarizes the different samples.

TABLE II. Specifications of iron contaminated samples of group A and B, respectively. Contamination level refers to the spin-on contaminated samples with medium (M), high (H), and very high (VH) contamination. Ion dose is given for ion implanted samples.

Sample Identification	Specific Resistance in Ωcm	Contamination Procedure	Contamination Level or Ion Dose in cm^{-2}
A2M, A2H	2	spin-on	M, H
A6M, A6H, A6VH	6	spin-on	M, H, VH
A20M, A20H, A20VH	20	spin-on	M, H, VH
A60M, A60H, A60VH	60	spin-on	M, H, VH
BL, BM, BH, BVH	10	ion implantation	$2.5 \cdot 10^{10}$, $1.0 \cdot 10^{11}$, $3.0 \cdot 10^{11}$, $1.0 \cdot 10^{12}$

Experimental

All Elymat and Elymos measurements were performed with a commercial Elymat II system (GeMeTec, Munich) which was adapted for the Elymos operation. The laser used in this study operates at a wavelength of 905 nm. The injection level measurements were conducted with increasing I_{Photo} with a minimum I_{Photo} of about 20 µA (low injection) and a maximum I_{Photo} of about 4800 µA (high injection condition). For very low lifetimes, the signal-to-noise (S/N) ratio was too low and the corresponding measurements were not evaluated. Always BPC measurements were performed before FPC mode operation in order not to deteriorate the wafer surface which would influence the BPC results.

Due to the inhomogeneous lateral iron contamination for samples of group A, a circular region in the center of the sample with a diameter of about 2 cm was chosen for lifetime evaluation. Within this region, each sample exhibited homogeneous lateral diffusion currents and thus iron concentrations. For group B samples, the implanted region was evaluated and corrected by the not iron related lifetimes from the rest of the sample. In the not implanted regions, however, the bulk lifetimes were much higher than in the implanted regions. Therefore the lifetimes in the implanted regions are predominantly affected by the iron contamination only.

Results and Discussion

BPC Measurement Results

Typical results of the BPC measurements for the samples of group A are shown in fig. 5. They are compared with simulations with distinct iron concentrations N_{Fe}. For low as well as for high I_{Photo}, a significant deviation between simulated and experimental lifetimes can be observed. For both injection level regimes experimental lifetimes are lower than simulated lifetimes. The deviations are more pronounced for higher lifetimes and can nearly be neglected for measurements with low injection lifetimes below 5 µs. All these observations are in good agreement with earlier results (5, 6).

The deviation for high injection conditions can partly be attributed to a non-ideal behavior of the samples which can be observed in the IV curve. Here, the diffusion current exhibits no saturation and even decreases after a local maximum. This is most pronounced for the samples with highest specific resistance (i.e., 60 Ωcm, see fig. 6). Therefore the correct saturated diffusion current would be slightly higher resulting in an increased Elymat lifetime. For specific resistances below 20 Ωcm, however, this effect could not be observed although even for these samples the deviation between simulated and measured lifetimes at high injection conditions exists.

At low injection conditions (i.e., low laser power I_{Photo}) the decrease in measured lifetimes with decreasing I_{Photo} can be explained by insufficient surface passivation using diluted HF as it is used in standard Elymat setups. Also Walz et al. (23) have stated that diluted HF might not fully suppress surface recombination during Elymat measurements. Frequently the work of Yablonovitch et al. (24) is cited to emphasize that HF should give the lowest surface recombination velocity for silicon samples (i.e., 0.25 cm/s). In that study, however, very high injection conditions were investigated which do not prevail for low I_{Photo} in Elymat measurements. In this work for the first time the Elymos method (5, 6) was successfully applied with native oxide as the insulating layer resulting in the expected behavior of low injection level lifetimes for iron contaminated samples (fig. 7). The native oxide grew during the storage of the sample in a wafer box without any specific cleaning or storage atmosphere. Even for high injection conditions, the electrostatic passivation seems to be superior to passivation with diluted HF for the sample A6M. Further studies will be performed on artificially processed very thin oxides without the use of high temperatures (i.e., oxidizing cleaning processes, ozone treatments) as the native oxides just grown during storage of the sample are not stable enough for repeating the measurements several times, especially after measurements at high injection. With the samples of group B, electrostatic passivation can be used using optimum thermal oxides.

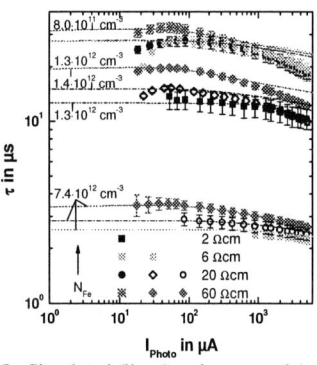

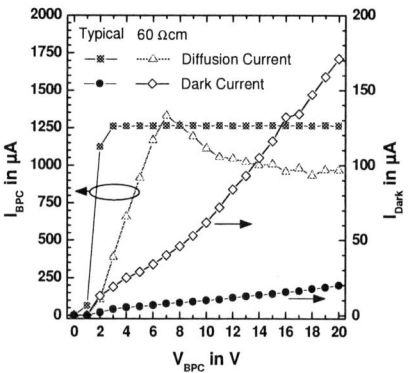

Figure 5. Simulated (lines) and measured (symbols) BPC lifetimes for different samples of group A. Iron concentrations N_{Fe} of the simulations are also given.

Figure 6. BPC IV curve for a typical sample and sample A60M at a photo current of 2600 µA.

The dependence of measured Elymat BPC carrier lifetimes on boron concentration is shown in fig. 8, where both, low injection Elymat lifetimes (symbols) and analytically calculated low injection lifetimes are compared. No BPC result for the sample with very high iron concentration and a specific resistance of 2 Ωcm could be determined due to very low S/N ratio. A good agreement between simulation and experiment over the whole investigated doping range can be observed for the different iron concentration levels (i.e., medium, high, and very high, see table II). Again this proves that the independently determined capture coefficients describe carrier recombination due to iron contamination correctly.

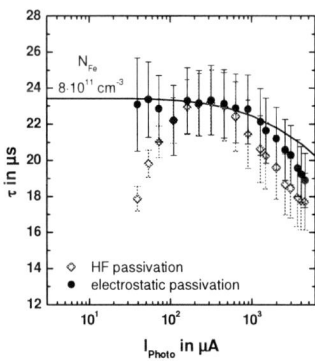 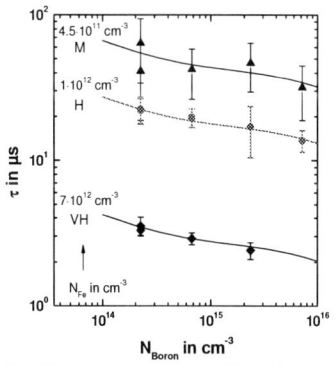

Figure 7. Comparison of both, results using the Elymos setup with electrostatic passivation and results from standard Elymat measurements using HF passivation for sample A6M.

Figure 8. Boron concentration dependence of low injection BPC carrier lifetimes for samples of group A with different iron concentrations. Lines represent calculated lifetimes using the model of Choo (14).

FPC Measurement Results

The interpretation of FPC measurements for samples with high carrier lifetimes is very challenging since the difference between I_{Photo} and I_{FPC} is very small. Therefore even the smallest measurement errors lead to large errors in the determination of the carrier lifetime (11). This is the main reason why FPC measurements are in practice used for monitoring or contamination control if BPC currents are too small or differentiation between bulk and surface near defects is important. Here, FPC results for samples of group A with high and very high contamination levels are presented and compared to DESSIS simulations with distinct iron concentrations (see figs. 9 and 10). The results are shown in two separate figures for the sake of clarity.

First of all, except of sample A2H for all samples the FPC lifetime increases with increasing I_{Photo}. This is in agreement with the simulations and has been observed by Polignano et al. (7) but there it was not explained but attributed to non-ideal sample characteristics or measurement settings. As it was discussed in the modeling section, however, this increase in Elymat FPC lifetime is real and due to an additional drift current whose fraction on the total current increases with increasing laser power. This effect is explained for the first time. For the sample A60H, the measured lifetimes strongly decrease for I_{Photo} higher than 1000 μA. This is a consequence of the effect that due to the high specific resistance of the sample the diffusion current does not reach its maximum for the maximum FPC voltage of 20 V (fig. 11). Therefore the correct FPC current would be much higher than the maximum detectable I_{FPC} resulting in much higher FPC lifetimes. Sample A2H does not show the predicted increase in FPC lifetimes with increasing I_{Photo}. In this case, this is due to the very high dark current of sample A2H (fig. 11) which leads to the non-ideal behavior of the diffusion current (i.e., decreasing diffusion current with increasing

V_{FPC}). For very high total currents, this effect is known and can be corrected using the approach proposed by Lippik (11). Dark current and total current during illumination are measured at the same V_{BPC} or V_{FPC}. The voltage drop across the SCR which is most important, however, is not the same for different currents due to unavoidable series resistances in the Elymat setup. This also holds for measurements with and without illumination leading to an erroneous determination of the diffusion current. The diffusion current is simply calculated as the difference between total current and dark current at the same V_{BPC} or V_{FPC}. Using the correction approach of Lippik, however, corrected FPC lifetimes of sample A2H are in agreement with simulated lifetimes (fig. 10).

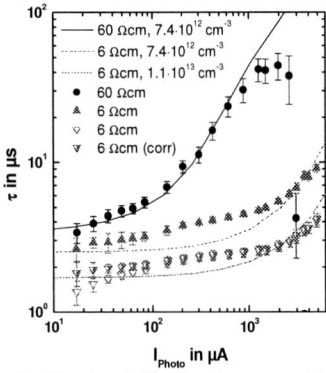 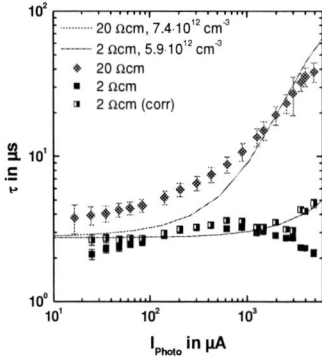

Figure 9. Simulated (lines) and measured (symbols) FPC lifetimes for samples A6H, A6VH, and A60H. The iron concentrations N_{Fe} for the simulations are also given.

Figure 10. Simulated (lines) and measured (symbols) FPC lifetimes for samples A2H and A20H. Iron concentrations N_{Fe} for the simulations and corrected data for sample A2H are also given.

Comparison of BPC and FPC Measurement Results

For samples of group A with high contamination levels, BPC and FPC results are compared (fig. 12). Only samples with sufficiently low lifetimes for FPC measurements and sufficiently high lifetimes to be able to perform BPC with adequate S/N ratio are presented. It is obvious that both, BPC and FPC measurements can be simulated in good agreement using the same iron concentrations for BPC and FPC simulations. In this context it is worth noting that the simulation for all three samples which nominally have the same iron concentration level, are described with the same iron concentration of $7.4 \cdot 10^{12}$ cm^{-3}. The results presented in fig. 12 clearly prove the validity of the modeling and the accuracy of the measurements.

Results for Iron Implanted Samples

Samples of group B where iron was implanted through a thermal silicon oxide of 45 nm thickness were prepared to exclude any surface recombination effects by enabling electrostatic passivation in Elymos mode with stable and reproducible conditions. The results of the BPC measurements for all implantation doses are shown in fig. 13. For all samples, a constant lifetime is measured even for the lowest photo currents. For samples BH and BVH the S/N ratio was too low for low photo currents. Only for high laser powers and lower iron concentrations simulations significantly overestimate BPC lifetimes. This effect is similar to the observations for samples of group A (fig. 5) and published data. Since surface recombination effects are most likely to be excluded and different values for the capture coefficients can not model this behavior it is assumed that this effect is probably due to trap assisted Auger (TAA) recombination (25). TAA is not considered for the simulations as corresponding capture coefficients are not known so far.

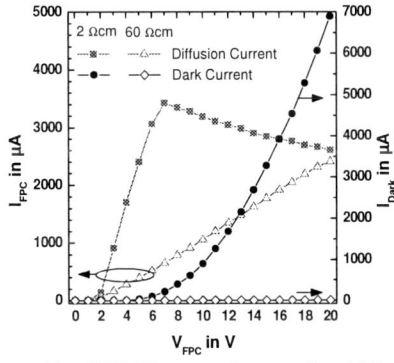 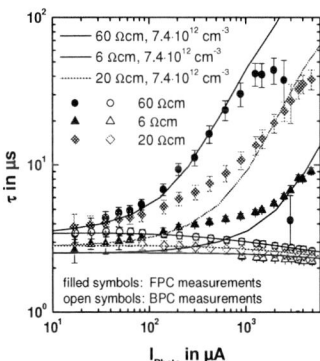

Figure 11. FPC IV curves for samples A2H and A60H for a photo current of 2600 μA.

Figure 12. Comparison of BPC and FPC lifetimes for samples A6H, A20H, and A60H. Iron concentrations N_{Fe} for the simulations (lines) are also given.

Simulated lifetimes in fig. 13 with iron concentrations adapted to describe the low injection lifetimes are represented by solid lines. Dashed lines show simulations which are used to estimate the accuracy of the adapted iron concentrations $N_{Fe,simul}$ which are compared to the iron concentrations $N_{Fe,implant}$ calculated from the nominal implanted ion dose assuming homogeneous iron concentrations throughout the sample thickness (fig. 14). From fig. 14 it can be observed that $N_{Fe,simul}$ values agree very well with $N_{Fe,implant}$ values for low iron concentrations and therefore high carrier lifetimes. This reflects both, the quality of the sample preparation as well as the high accuracy of the Elymos measurements since especially high carrier lifetimes are affected by a huge variety of effects and parameters. That $N_{Fe,simul}$ underestimates $N_{Fe,implant}$ for high iron concentrations might be due to insufficient diffusion time or temperature which would become more obvious for higher iron concentrations.

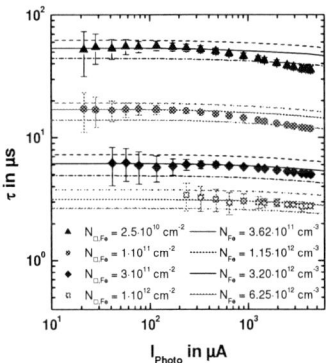 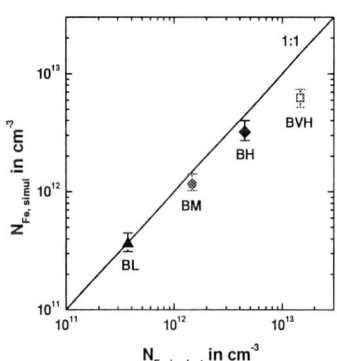

Figure 13. Comparison of simulated (lines) and measured (symbols) BPC lifetimes for samples BL, BM, BH, and BVH. Electrostatic passivation is used.

Figure 14. Simulated iron concentrations $N_{Fe,simul}$ vs. iron concentrations $N_{Fe,implant}$ calculated from implantation doses for samples shown in fig. 13.

Summary

In this work, the Elymat carrier lifetime method is used for an extensive study of the injection level dependence of carrier lifetime for intentionally iron contaminated boron-doped silicon. For the first time, a consistent modeling and interpretation of results from both, FPC and BPC Elymat measurement modes for different iron and boron concentrations is presented. Apart from the improved simulation, optimized measurement conditions and evaluation procedures are proposed and discussed in detail. Deviations between measured and theoretically simulated BPC lifetimes for low injection conditions are shown to be due to insufficient surface passivation using standard passivation with diluted HF. Applying electrostatic passivation with the Elymos method, however, allows the determination of correct lifetimes. Even native oxide can be used for electrostatic passivation. The results clearly prove that the Elymat method is suited for advanced carrier lifetime analysis allowing the identification and quantitative determination of iron in case of iron being the relevant contaminant in boron doped silicon.

References

1. D.V. Lang, *J. Appl. Phys.*, **45**, 3023 (1974).
2. G. Ferenczi, T. Pavelka, and P. Tüttö, *Jpn. J. Appl. Phys.*, **30**, 3630 (1991).
3. S. Rein, *Lifetime Spectroscopy, Springer Series in Material Sciences*, **85** (2005).
4. V. Lehmann and H. Föll, *J. Electrochem. Soc.*, **135**, 2831 (1988).
5. M.L. Polignano et al., in *Optical Characterization Techniques for High-Performance Microelectronic Device Manufacturing II*, SPIE Proc. **2638**, 14 (1995).
6. D. Walz, J.-P. Joly, and G. Kamarinos, *Appl. Phys. A*, **62**, 345 (1996).
7. M. L. Polignano et al., *Mater. Sci. Eng. B*, **55**, 21 (1998).
8. M.L. Polignano et al., in *Analytical and Diagnostic Techniques for Semiconductor Materials, Devices, and Processes*, ECS PV **99-16**, p. 38 (1999).
9. M. Rommel et al., in *Crystalline Defects and Contamination: Their Impact and Control in Device Manufacturing IV*, ECS PV **2005-10**, p. 113 (2005).
10. S. M. Sze, *Physics of Semiconductor Devices*, 2nd ed., Wiley, New York (1981).
11. W. Lippik, *Further Development of the Elymat Technique* (PhD thesis, in German), University of Kiel, Germany (1996).
12. DESSIS 10.0, Synopsis Inc., Mountain View, CA (2004).
13. M. Rommel et al., *Diff. Defect Data B*, **82-84**, 373 (2002).
14. S. C. Choo, *Phys. Rev. B*, **1**, 687 (1970).
15. M. Kerr and A. Cuevas, *J. Appl. Phys.*, **91**, 2473 (2002).
16. H. Daio et al., in *18th International Conference on Defects in Semiconductors*, Mat. Sci. Forum *196-201*, p. 1817 (1995).
17. D. Gilmore et al., *J. Electrochem. Soc.*, **145**, 621 (1998).
18. A. Kempf et al., in *Advanced Workshop on Silicon Recombination Lifetime Characterization Methods*, ASTM STP **1340**, p. 259 (1998).
19. H. Park, C.R. Helms, and D. Ko, *J. Electrochem. Soc.*, **145**, 1724 (1998).
20. W. Shockley and W. Read, *Phys. Rev.* **87**, 835 (1952).
21. R. Hall, *Phys. Rev.* **87**, 387 (1952).
22. M. Hourai et al., *Jpn. J. Appl. Phys.*, Part 2 **27**, L2361 (1988).
23. D. Walz et al., *Sem. Sci. Techn.* **10**, 1022 (1995).
24. E. Yablonovitch et al., *Phys. Rev. Lett.* **57**, 249 (1986).
25. A. Haug, *Phys. Status Solidi b*, **108**, 443 (1981).

CHAPTER 4

ELECTRICAL AND SCANNING PROBE MICROSCOPY TECHNIQUES

ECS Transactions, 10 (1) 129-140 (2007)
10.1149/1.2773983 ©The Electrochemical Society

Structural and Electrical Characterization of Dielectrics, Carbon Nanotubes and Nanoelectronic Devices by Means of Scanning Probe Microscopy

Udo Schwalke

Institute for Semiconductor Technology & Nanoelectronics,
Darmstadt University of Technology,
64289 Darmstadt, Germany

During the past years, atomic force microscopy (AFM) has become an indispensable tool for non-destructive characterization of structures at the nanometer-scale. By implementing current sensing techniques, the topographic imaging capabilities of the AFM have been extended. Conductive atomic force microscopy (C-AFM) allows direct correlations of feature locations with the corresponding electrical properties. In this paper, several examples on the application of combined AFM and C-AFM techniques will be presented in order to illustrate the potential of the scanning probe microscopy (SPM) technique for structural and electrical characterization of gate dielectrics, carbon nanotubes as well as nanoelectronic device structures.

Introduction

Advanced device fabrication requires numerous steps for process control to ensure the desired functionality. Process metrology includes control of geometrical dimensions as well as materials and device characterization (1). However, most of the common characterization techniques are either destructive or do not provide the required lateral resolution and laborious sample preparation is often needed. This situation is further aggravated due to the ongoing aggressive down-scaling. Microelectronics has entered the nanoelectronics era and state-of-the-art silicon CMOS technologies are utilizing sub-100 nm feature sizes. This continuous top-down miniaturization of silicon-based nanoelectronics is expected to continue into the sub-10 nm range (2). Furthermore, for the post-silicon CMOS era very promising bottom-up concepts have been proposed utilizing carbon nanotube (CNT) molecular electronic devices (3) and circuits (4).

The metrology of these nanometer-scale structures is very challenging and demands for powerful and cost-efficient diagnostic techniques. Since the invention of the scanning tunneling microscope (STM) by Binnig and Rohrer (5) in 1981 this situation has largely improved and meanwhile a versatile family of scanning probe microscopy (SPM) techniques has evolved (6). The atomic force microscopy (AFM) is a further development of the STM (7). However, unlike the STM, the AFM technique is much more versatile and can image insulating materials. More recently, the topographic imaging capabilities of the AFM have been extended by implementing current sensing techniques (8). These so called conductive AFM (C-AFM) or tunneling AFM (TUNA) systems allow a direct correlation of a feature location precisely with its electrical signature because topography information and electrical data are recorded simultaneously.

In this paper we will present several applications of AFM and C-AFM techniques which have been used successfully for structural and electrical characterization of gate dielectrics, carbon nanotubes as well as nanoelectronic devices. The potential of this simple, non-destructive characterization technique will be illustrated and the impact on process development and fabrication of carbon-nanotube field-effect transistors (CNTFETs) will be demonstrated.

AFM and C-AFM Instrumentation

The AFM base-system used in this work is a Dimension 3100 Scanning Probe Microscope supplied by Veeco Instruments (9). As in most AFMs used today, an inexpensive optical levering technique is used to measure the cantilever deflection as illustrated in Fig. 1. A well collimated laser beam is reflected on the cantilever and projected on a position-sensitive photo-detector. For topographic measurements basically three different operating modes (1, 6) can be used: In the 'contact mode' the repulsive part of the surface potential is probed and the main feedback loop is programmed to increase the tip-sample distance when the detected force increases. The 'non-contact mode' refers to the operation in the attractive part of the interaction potential which is used to detect small attractive forces for dynamic force microscopy. Finally, in the 'tapping mode' the tip penetrates temporary into the repulsive part of the potential during its oscillation cycle.

An additional current sensing module attached to the D3100 SPM allows performing electrical current measurements simultaneously with the topographical ones. The electrically conductive probe is scanned in contact with the sample which is biased through the chuck. The resulting current between the sample and the conductive probe is measured to generate a current image. The bias can be supplied either to the wafer chuck

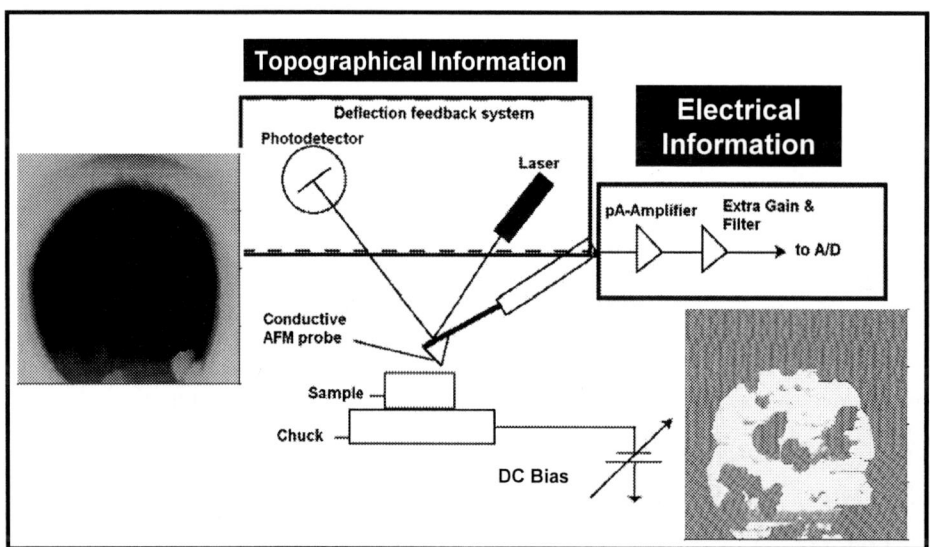

Figure 1. Schematic of the AFM / C-AFM measurement setup (center). Topographical AFM scan (left inset) of etched contact hole and related electrical C-AFM image (right inset) are obtained simultaneously. The bright areas indicate electrical contact with silicon and complete removal of the insulator.

(i.e. the wafer back side) or to specially designed connecting pads on the wafer surface to allow for more sophisticated device measurements. A simple illustrative example for process control is shown in the insets of Fig. 1. The oxide removal across the bottom of the contact hole after reactive-ion-etching (RIE) can be easily verified by the C-AFM current image simultaneously obtained with the topographical AFM scan. The bright areas indicate enhanced currents. When taking into account the tip area of ~100 nm^2, the measured currents of typically around 5 pA translate into a high current densities of approximately 5 A/cm^2 indicating that either no dielectric or only a native oxide covers the etched bottom of the contact hole. In addition to the imaging mode, current-voltage characteristics may be obtained locally by keeping the tip in a fixed lateral position during the contact mode. The probe tip radius of around 10 nm and the current sensitivity in the fA range allow high resolution electrical analysis.

Electrical Characterization of High-K Gate Dielectrics

The ITRS roadmap (10) predicts that gate oxide thickness has to be scaled down aggressively towards the sub-nanometer range within the next decade. Alternative high-k gate dielectrics like HfO$_2$ (11) are currently investigated as replacement for thermally grown silicon dioxide (SiO$_2$) in an attempt to overcome the severe gate tunneling currents. Recently it was shown that rare earth high-k dielectrics epitaxially grown on the Si-surface by molecular-beam-epitaxy (MBE) have excellent dielectric properties (12). Crystalline praseodymium oxide (Pr$_2$O$_3$) films (13 - 15) and gadolinium oxide (Gd$_2$O$_3$) films (16, 17) have been successfully integrated in CMOS technologies. However, enhanced charge trapping susceptibility and high defect density is still an issue (18, 19).

Usually macroscopic devices like MOSFETs or capacitor test structures as shown in Fig. 2 (left) are used for leakage current measurements on gate dielectrics. However, due to the macroscopic gate electrode, different leakage mechanisms in one and the same film are often superimposed and it is complicated, if not impossible, to separate each one of them from the rest by macroscopic measurements. The macroscopic gate electrode will

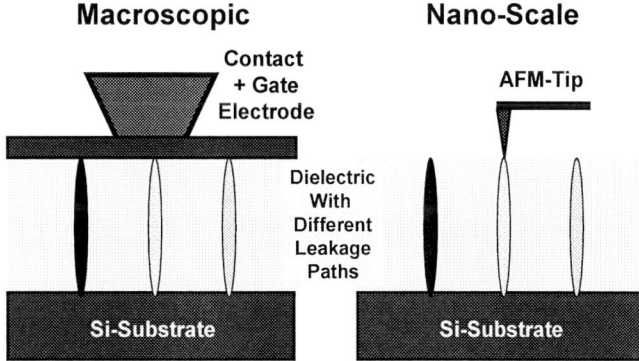

Figure 2. Electrical characterization of gate dielectrics illustrated for a macroscopic device (left) and at the nanometer-scale with C-AFM (right). Different leakage current paths are averaged by the macroscopic gate electrode. With C-AFM the individual leakage paths can be separated and the current transport mechanisms investigated.

Topographical Scan   Current Image

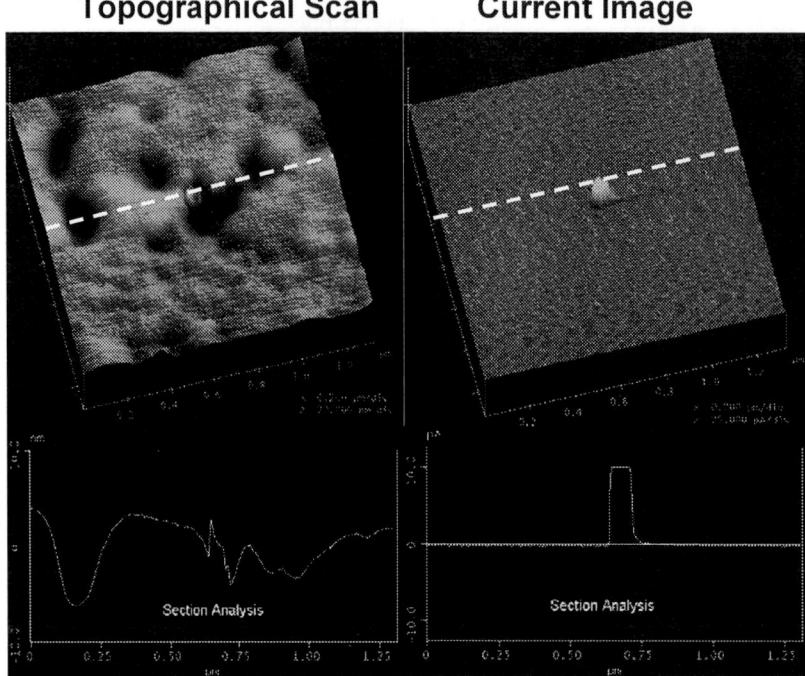

Figure 3. Topographical (left) and current (right) scans of crystalline Pr_2O_3 high-k dielectric with a physical thickness of 19 nm. The structural defect is clearly visible in the current image (right).

simply integrate over the various leakage-current paths. Conductive AFM on the other hand allows electrical measurements with nanometer-resolution in which individual leakage spots can be discriminated and their electrical behavior can be studied separately as illustrated in Fig. 2 (right). In this case, the C-AFM probe tip itself acts as a gate electrode and scans across the dielectric surface. With its high spatial resolution and the additional topographic information, C-AFM provides a better understanding on the microscopic origin of the observed complex macroscopic electrical behavior in high-k dielectric films.

In Fig. 3 an example of a structural defect in a Pr_2O_3 film with a physical thickness of 19 nm is shown. Due to the rough surface, structural defects are difficult to find in the topographical image. However, structural defects often lead to excessive leakage currents and can be therefore easily identified by C-AFM current scans as shown in Fig. 3 (right). Even for the low bias of 2 V, a very high ohmic leakage current above 10 pA (i.e. > 10 A/cm^2) is measured. Interestingly, despite the corrugated surface, the leakage current distribution is very uniform and at a low level of approximately 200 fA, even in areas where the thickness is drastically reduced by 40 %.

In the case of Al/TiO_2-SiO_2/Si capacitors contradictive results on leakage current characteristics have been obtained from macroscopic device measurements (19). AFM measurements reveal that two different kinds of structural defects occur within the dual

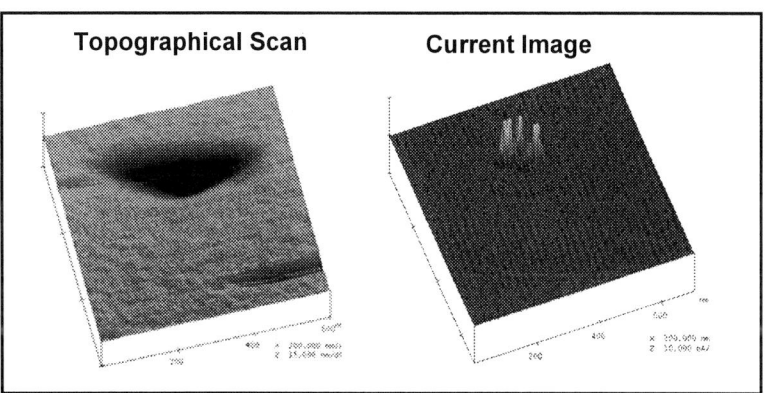

Figure 4. Topographical scan (left) and current image (right) of a dual-layer (TiO$_2$-SiO$_2$) high-k dielectric taken after anneal at 950°C.

layer TiO$_2$-SiO$_2$ high-k gate dielectric. After annealing the TiO$_2$-SiO$_2$ samples in inert ambient at 950°C, local thinning is observed as shown in Fig. 4. However, in contrast to the crystalline epitaxially-grown high-k films, enhanced leakage currents are observed by C-AFM in this case. We suspect that Ti may diffuse laterally, leaving a thinned TiO$_2$ on the interfacial SiO$_2$ buffer-layer as illustrated in Fig. 5. The size of these leaky spots is up to a few hundred nanometers in diameter and the spot density is approximately 10^4cm^{-2}. In order to measure the current-voltage (I-V) characteristics via C-AFM, the probe tip is held in a fixed position in contact at the center of the defect. Nearly symmetrical current-voltage dependence is observed as evident from Fig. 5 which supports the proposed lateral migration of titanium. The enhanced leakage current observed in these cases is probably due to direct-tunneling or F-N tunneling primarily through the SiO$_2$ buffer-layer which is still present.

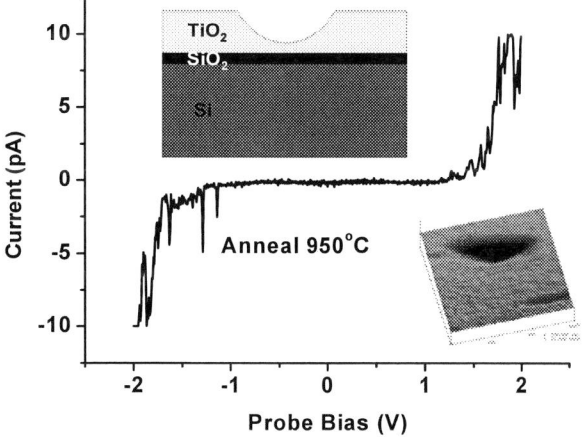

Figure 5. C-AFM current-voltage plot recorded with tip in a fixed position and in contact with the defect shown in the inset (lower right). Symmetric leakage is due to tunneling. Cross-section of the proposed defect structure is shown in the inset (top).

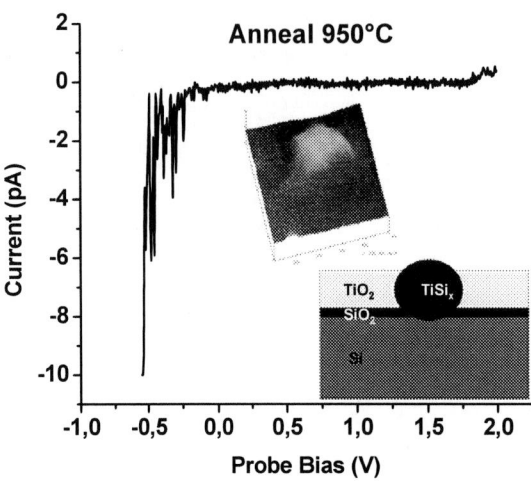

Figure 6. Schottky-diode I-V characteristics recorded with C-AFM in the point-contact mode on the hillock-like defect shown in the inset (center). Cross-section of the proposed defect structure is shown in the inset at the bottom.

A second type of structural defects, hillock-like defects, has been observed in addition, as shown in Fig. 6. We suspect that the hillocks are the result of silicide formation, i.e. $TiSi_x$. During annealing at 950°C, titanium may diffuse vertically all the way down to the silicon at some spots where the SiO_2 buffer-layer is incomplete. The silicide forms a metal-semiconductor contact causing the Schottky-diode-like current-voltage characteristics displayed in Fig. 6. By means of C-AFM quite different microscopic leakage current mechanisms originating from structural defects have been identified which are responsible for high leakage observed in macroscopic devices (19).

Carbon Nanotube Growth and CNTFET Characterization

Carbon nanotubes (CNTs) are promising candidates to replace Si-CMOS in future nanoelectronics because of their unique one-dimensional geometry giving them excellent carrier transport properties. Since 1998 (20, 21), it is known that semiconducting single-walled carbon nanotubes (s-SWNTs) can act as the channel in carbon-nanotube field-effect transistors (CNTFETs). Currently CNTFETs are fabricated and investigated by many research groups. However, the fabrication processes used are often complicated, including both separate growth (22) and manual manipulation of the CNTs (20, 21) or require multi-step lithography with the risk of misalignment. Obviously, large scale integration remains a significant challenge to the realization of CNT-based nanoelectronics.

At our institute we have developed a novel process to overcome the limitations of the manual fabrication of CNTFETs (23 - 26). The process is based on chemical-vapor-deposition (CVD) growth of CNTs using a thin aluminum/nickel 'sacrificial' catalyst (1 nm Ni on 10 nm Al) which transforms itself after CNT growth into a high-k dielectric (i.e.

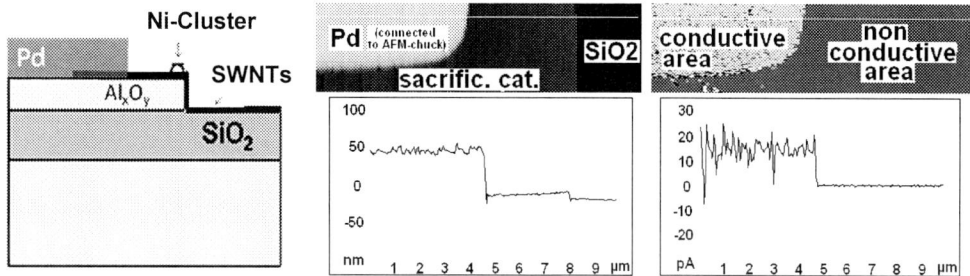

Figure 7. Schematic cross-section (left) of the test structure used for AFM and C-AFM measurements. AFM height scan (center) with cross-section along the white line shows the step created by the sacrificial catalyst on SiO_2. However, only the Pd-electrode is visible in the C-AFM current image (right) which is simultaneously obtained with the topographical image. Both, sacrificial catalyst area and SiO_2 are non-conductive as clearly evident from the current cross-section along the white line.

Al_xO_y) covered with dispersed Ni-nanoclusters (25, 26) as shown in Fig. 7. SWNTs are grown uniformly across the wafer surface and subsequently contacted with palladium for S/D contacts and the Si-substrate acts as a gate electrode (see also Fig. 11, left). The process contains neither complicated manipulations of the SWNTs nor multi-step lithography, avoiding the risk of misalignment. We choose the in-situ growth method because it appears the most practical approach for future use in high-volume fabrication of advanced nanoelectronics where millions of transistors need to be integrated per chip at low cost.

Carbon Nanotube Growth: Characterization and Optimization

Prior to the fabrication of CNTFETs, the above mentioned novel process has to be verified, i.e. the transformation of the metallic sacrificial catalyst into the Al_2O_3-like high-k dielectric. In order to perform the necessary C-AFM measurement after CVD, a test structure was designed (Fig. 7, left) which allows pre-structuring the metallic catalyst (Al/Ni) after metal deposition. The metal layer creates a visible step on the SiO_2 which is still present in the topographical scan (Fig. 7, center) after CVD. Subsequently to CVD, Pd-electrodes are deposited and electrically connected with the chuck to allow for C-AFM measurements. As a result, two steps (i.e. Pd on catalyst and catalyst on SiO_2) are geometrically visible as seen on the AFM height scan of Fig. 7 (center). However, in the C-AFM current image the catalyst cannot be distinguished from the SiO_2 as evident from Fig. 7 (right). Only the Pd electrode is visible because it is conductive and increases the current. Obviously, the ultra-thin aluminum film turns into an insulator during CVD by reaction with SiO_2. To which extent some additional reaction with adsorbed moisture from ambient air may also stimulate the conversion from Al into Al_xO_y is currently investigated. Taking advantage of the C-AFM, the conversion from the metallic into an insulating material is confirmed, as evident from Fig. 7. The sacrificial catalyst will not short-circuit the SWNT devices and will allow the gate field to penetrate to control device characteristics of CNTFETs (cf. Fig. 11).

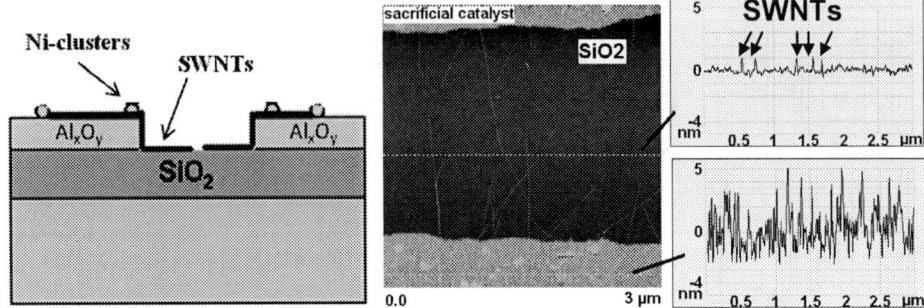

Figure 8. Topographical AFM measurements: schematic of the test structure (left) used for height scans (center). Corresponding cross-sections (right) showing the different surface roughness of SiO_2 (RMS < 0.5 nm) and sacrificial catalyst (RMS > 3 nm). Only on the smooth SiO_2 the SWNTs with a diameter of approximately 1 nm can be identified.

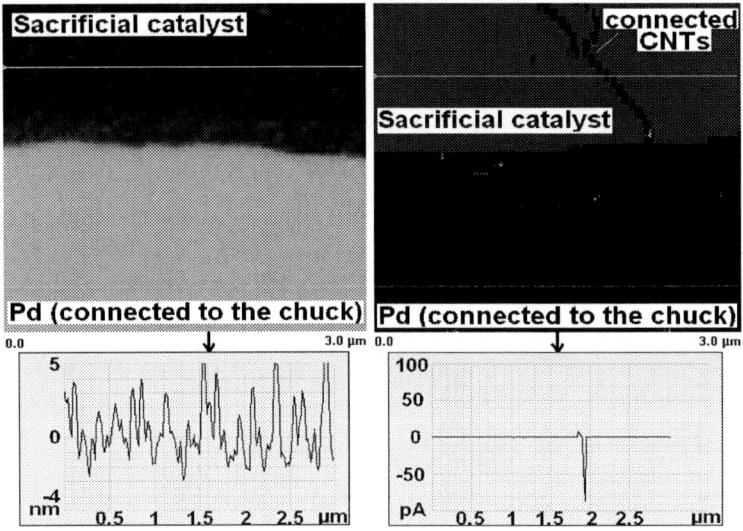

Figure 9. Comparison of topographical and conductive AFM measurements on sacrificial catalyst. Due to the high roughness of the sacrificial catalyst no SWNTs are detectable by topographical AFM (left). Using C-AFM (right) the existence of SWNTs can be verified electrically.

In addition, the selected "sacrificial" catalyst should be appropriate to stimulate selective growth of SWNTs. Topographical AFM has been extensively used to achieve results on the selective growth of SWNTs on pre-structured catalyst test structures shown in Fig. 8. The role of nickel as catalyst particle to stimulate SWNT growth is evident from the AFM hight scan (tapping mode) which indicates that SWNTs always start to grow from a Ni-cluster and extend onto the SiO_2. SWNTs with a diameter of approximately 1 nm are clearly detectable on the smooth thermally grown SiO_2 with a RMS roughness of < 0.5 nm by means of topographical AFM shown in Fig. 8 (right). This result confirms that the selective growth of SWNTs is in fact achieved and that the presence of multi-walled carbon nanotubes (MWNT) which always have a diameter above 3 nm can be excluded. However, due the high surface roughness of the sacrificial catalyst it is impossible to verify with AFM the presence of SWNTs on the sacrificial catalyst. This drawback of AFM is omitted when applying the conductive AFM method instead. In order to perform electrical C-AFM measurements, palladium contacts are deposited on the sacrificial catalyst area and are electrically connected to the AFM chuck. When an electrical path between tip and chuck through SWNTs is created, the SWNTs can be identified electrically in the C-AFM current image as shown in Fig. 9 (right). Taking advantage of C-AFM, restrictions of the topographical AFM can be resolved.

Fabrication and Characterization of CNTFETs

Based on these results, an extremely simple process for the fabrication of CNTFETs has been proposed and was realized (25, 26). Since the SWNTs are grown uniformly across the wafer surface covered with the sacrificial catalyst, palladium S/D contacts can be placed on any desired location and the Si-substrate may be used as a gate electrode. The process contains neither complicated manipulations of the SWNTs nor multi-step lithography, avoiding the risk of misalignment. Also, after having connected S/D contacts to the AFM chuck, more detailed electrical information at the nanoscale can be obtained. An example is given in Fig. 10 in which we have made a first attempt to perform resistance measurements on individual SWNTs by C-AFM. In this case, the probe tip stays at a fixed position in contact with the SWNT and a bias ramp is applied, so that the current- voltage (I-V) characteristic is obtained, from which the total resistance between the probe tip and the chuck can be calculated. The fairly high value of 80 MΩ results probably in part from the high intrinsic resistance of the SWNT that should be semiconducting and being in the off-state since no appropriate gate bias is applied to turn it on completely. (Note that a metallic SWNT is always conducting). The value for the s-SWNT agrees well with results from Bachtold et al. (27) of 60 MΩ obtained on semiconducting SWNTs measured under similar conditions. Additional contributions to the total resistance are due to contact resistances of the CNT to the probe tip and the S/D electrodes as well. In fact, we used a probe tip coated with platinum on chromium and Pt is known to lead to poor metal-CNT contact because of the large Schottky-barrier formed at the metal-CNT junction (28).

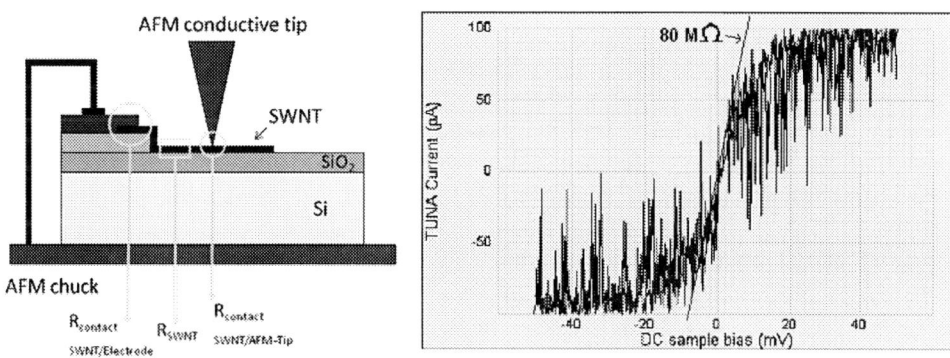

Figure 10. Schematic representation of measurement set-up with resistive components (left). Example of current-voltage characteristics obtained by C-AFM measurement in at the device level (right).

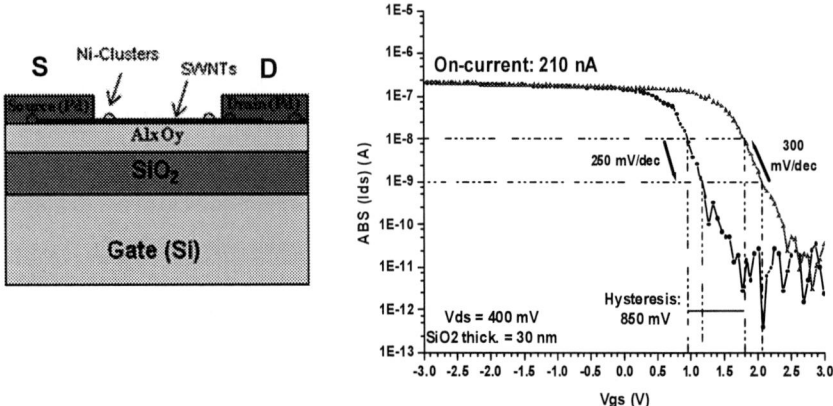

Figure 11. Schematic cross-section of CNTFET (left) and measured sub-threshold characteristics of a PMOS-like CNTFET (right).

The application of the AFM and C-AFMs technique for process optimization resulted in fully functional CNTFETs with excellent macroscopic device characteristics. In Fig. 11 an example of the measured macroscopic device characteristics is shown. The transistor is unipolar and PMOS-like, i.e. a negative gate bias is required for turn-on. Similar to conventional MOSFETs, the CNTFETs show a strong dependence on oxide thickness, i.e. with decreasing gate oxide thickness the sub-threshold characteristics of the CNTFETs and current-drive improves (26). The on/off current ratio is in the 10^5 range and comparable with the values of previously published manually fabricated CNTFETs. Note that the device shown in Fig. 11 contains only one SWNT so that the on-current corresponds to 150 μA/μm width at $V_{ds} = 0.4$ V for a channel length of 2.5 μm, already exceeding the current-drive requirements of the IRTS roadmap (10) when scaling down this device to 45 nm. Furthermore, the presented simple CNTFET fabrication process is CMOS-compatible and opens the possibility to realize hybrid CNT-CMOS circuits.

Conclusion

Applications of the AFM and C-AFM method have been presented to illustrate the potential of scanning probe microscopy techniques in semiconductor characterization. The AFM by itself is already a powerful technique, being capable of imaging topographies at very high resolution both vertically and laterally. However, the identification of nanometer size objects is restricted to smooth surfaces, like in the case of CNTs on SiO_2. It has been shown, that by implementing current-imaging capabilities via C-AFM this limitation can be resolved. The C-AFM extension is very simple but nevertheless it offers a valuable additional degree of freedom in nanometer scale characterization. Individual nanoscale devices can be probed electrically as well as helpful information on dielectrics or silicon front-end processing is obtained. For the future one of the most important assets of the AFM is probably its adaptability to various additional measurement needs. Additional derivative techniques together with novel implementations of semiconductor material and nanoelectronic device characterization are expected to be developed in the future.

Acknowledgments

The author is indebted to L. Rispal, Y. Stefanov and R. Endres who have largely contributed to the results presented in this article. Also, the support of F. Wessely and F. Zaunert is greatly acknowledged. The author would also like to thank the entire staff of the IST for wafer processing, in particular, K. Haberle, R. Heller, G. Hess and G. Tzschöckel. Part of the work was funded by the German Federal Ministry of Education and Research (BMBF) under the MEGA EPOS project (13N9259) and the KrisMOS project (01M3142C) as well as the German Research Foundation (DFG) under the StraMNano Project (SPP 1159).

References

1. D. K. Schroder, *Semiconductor Material and Device Characterization*, Third Edition, Wiley-Interscience, New York (2006).
2. B. Doris et al., *IEEE International Electron Devices Meeting Technical Digest (IEDM)*, 267 (2002).
3. R. Martel, T. Schmidt, H. R Shea, T.Hertel, Ph. Avouris, *Appl. Phys. Lett., 73*, 2447 (1998).
4. A. Bachtold, P. Hadley, T. Nakanishi, C. Dekker, *Science, 294*, 1317-1320 (2001).
5. G. Binnig, H. Rohrer, C. Gerber, and E.Weibel, *Physical Review Letters*, 49, 57 (1982).
6. S. Kalinin A. Gruverman, *Scanning Probe Microscopy*, Springer, New York (2007)
7. G. Binnig, C. F. Quate, and C. Gerber, *Physical Review Letters,* 56, 930 (1986).
8. A. Olbrich, B. Ebersberger, C. Boit, *Appl. Phys. Lett., 73*, 3114 (1998).
9. http://www.veeco.com/products/details.php?cat=1&sub=1&pid=178
10. http://www.itrs.net/Links/2006Update/2006UpdateFinal.htm
11. A. Kerber, E. Cartier, L. Pantisano, R. Degraeve, T. Kauerauf, Y. Kim, A. Hou, G. Groeseneken, H.E. Maes, U. Schwalke, *IEEE Electron Device Letters*, 24, 87 (2003).
12. U. Schwalke, K. Boye, K. Haberle, R. Heller, G. Hess, G. Müller, T. Ruland, G. Tzschöckel J. Osten, A. Fissel, H.-J. Müssig, *Proceedings of the 32nd European Solid State Device Research Conference (ESSDERC),* 407 (2002).
13. U. Schwalke, Y. Stefanov, R. Komaragiri, T. Ruland, *Proceedings of the 33rd European Solid State Device Research Conference (ESSDERC)*, 247 (2003).
14. H.D.B. Gottlob et al., *Journal of Non-Crystalline Solids*, **351**, 1885 (2005).
15. Y. Stefanov, R. Komaragiri, U. Schwalke, *Proceedings of the International Conference on Memory Technology and Design (ICMDT)*, 167, Giens, France, (2005).
16. R. Endres, Y. Stefanov, U. Schwalke, *ECS Transactions*, **3**, (2), 297 (2006)
17. R. Endres, Y. Stefanov and U. Schwalke, *Microelectronics Reliability*, **47**, 528 (2007)
18. U. Schwalke, Y. Stefanov, *Microelectronics Reliability*, **45**, pp. 790 – 793, 2005

19. Y.Stefanov et al., *The Electrochemical Society Conference on Crystalline Defects and Contamination (ECS-DECON)*, Grenoble, France, (2005).
20. R. Martel, T. Schmidt, H. R. Shea, T. Hertel, Ph. Avouris, *Appl. Phys. Lett.,* **73**, 2447, (1998).
21. A. Bezryadin, A.R.M. Verschueren, S.J. Tans, C. Dekker, *Phys. Rev. Lett.,* **80**, 4036 (1998).
22. A. Barreiro, D. Selbman, T. Pichler, K. Biedermann, T. Gemming, M.H. Ruemmeli, U. Schwalke, B. Buechner, *Applied Physics A*, **82** (4), 719 (2006)
23. L. Rispal, Y. Stefanov, F. Wessely, U. Schwalke, *Japanese Journal of Applied Physics*, **45**, 3672 (2006).
24. L. Rispal, T. Ruland, Y. Stefanov, F. Wessely, U. Schwalke, *ECS Transactions,* **3**, (2) 441 (2006)
25. L. Rispal, U. Schwalke, "Fabrication-Process for CNTFETs Based on Sacrificial Catalyst: Device Characterization and Conductive-AFM Measurements", Nanotech Northern Europe 2007 (NTNE 2007), Helsinki, Finland (2007).
26. L. Rispal, R. Heller, G. Hess, G. Tzschöckel, U. Schwalke, *ECS Transactions,* **11**, in print, (2007)
27. A. Bachtold et al., *Physical Review Letters*, **84**, 6082 (2000).
28. M.S. Kabir, R. E. Morjan, O. A. Nerushev, P. Lundgren, S. Bengtsson, P. Enokson, E. E. B Campbell, *Nanotechnology,* **16**, 458 (2005).

Methods for the Controlled Reduction of Carrier Lifetime in Power Devices

F. Hille, F.-J. Niedernostheide, H.-J. Schulze

Infineon Technologies AG, Am Campeon 1-12, DE-85579 Neubiberg, Germany

> Power devices require a thick lightly doped drift layer to support the required blocking capability. In order to obtain also an optimized on-state and dynamic switching behavior, the shape of excess charge carrier profile in the drift region has to be tailored. Often, and especially for freewheeling diodes, this can only be achieved by a precise adjustment of the carrier lifetime. In this work, we use the free carrier absorption technique to determine the excess charge carrier profile in power diodes. Methods to extract the carrier lifetime from the carrier profile of platinum diffused diodes are demonstrated. The injection level and temperature dependence of the carrier lifetime are determined. The origin of the specific shape of the excess charge carrier profile is analyzed by complementary measurements of the deep level profile by DLTS.

Introduction

High-power devices, such as IGBTs, freewheeling diodes or thyristors, require a lightly doped drift layer to achieve the necessary blocking capability. Usually n-type float-zone silicon with a high resistivity is used as starting material. Since these devices also have to be able to switch currents of several kA, the active area of the devices ranges from a few mm² to the total area of 5-inch or 6-inch wafers. In contrast to integrated circuit, where only a surface part of the wafer is used as active region, the high power densities can only be managed if the whole volume of the chips is used as active volume. Device fabrication therefore requires a processing of the both sides of the wafer and the current transport is in vertical direction. To achieve low forward voltages, bipolar current transport is used. An electron-hole plasma floods the lightly doped drift region in order to decrease the resistivity by several orders of magnitude. Typical applications of power devices require the change from conducting to blocking states at switching frequencies up to several 10 kHz. The removal of the excess charge during the switching transients gives rise to a significant amount of switching losses. Tailoring the switching behavior requires the shaping of the excess charge carrier profile. Proper design of the gate and emitter structures on both sides of the wafer control the amount of charge injected from the surface regions into the volume. Further shaping of the profile within the volume requires the adjustment of the carrier lifetime. For this purpose, doping with transition metals like platinum or gold [e.g. (1-5)], which reduces the carrier lifetime in the whole wafer, or irradiation with helium ions or protons [e.g. (6-8)], which only affect a certain volume of the wafer, is commonly used. All these methods lead to deep trap levels within the band-gap and enhance the recombination. Although a number of investigations have been carried out to determine the physical properties of the recombination centers, the characterization and understanding of lifetime dependence is still incomplete (9).

In this work, we exploit the Free Carrier Absorption (FCA) technique to determine the excess charge carrier profile of platinum diffused freewheeling diodes. The carrier lifetime is extracted from the curvature of the carrier profile. Analyzing diodes subjected to different platinum processes, the dependence of the carrier lifetime on platinum diffusion temperature is determined. Furthermore, the carrier lifetime dependence on the injection level and the operating temperature are extracted. The deviations of the results obtained for diodes of 650 µm und 110 µm thickness lead to the hypothesis, that the platinum is distributed within the diode according to a *U*-shaped profile and that only part of this profile is present for the thinner diode. Complementary, spatially resolved Deep Level Transient Spectroscopy (DLTS) measurements are carried out to determine the trap level distribution within the diode and confirm our hypothesis.

Free Carrier Absorption Technique

The FCA technique is a powerful tool to measure the excess charge carrier profile within vertical power devices. The schematic setup is shown in figure 1.

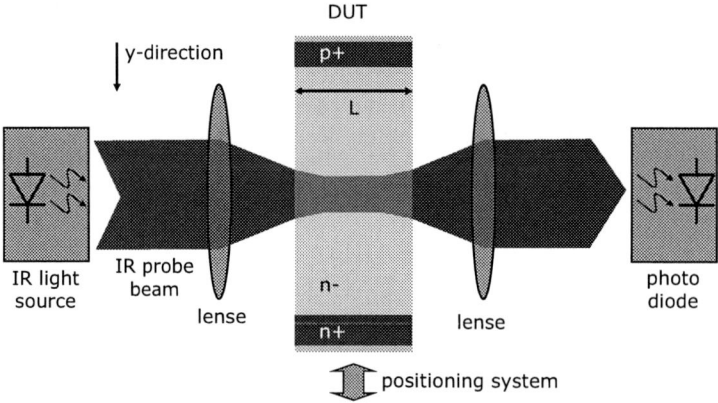

Figure 1. Schematic of the experimental setup for free carrier absorption measurements.

An infrared light beam is focused into the Device Under Test (DUT) through its previously polished side facets and the transmitted light intensity is detected by a photo diode. A laser diode with a wavelength of 1.55 µm and a very short coherence length of 20 µm is chosen as light source. At this wavelength, optical generation of electron-hole pairs in the silicon sample is avoided and the light beam solely acts as probing beam. As the DUT, in this case a vertical power diode, is turned on by a forward current pulse, holes are injected from the p$^+$ anode region (top side of the DUT) and electrons are injected from the n$^+$ cathode side (bottom side of the DUT). As a consequence, an electron-hole-plasma floods the n$^-$ drift layer of the diode. Due to the very short coherence length of the probing beam, Fabry-Perot interferences between the polished side facets of the sample are suppressed. Therefore, the free carrier absorption by the injected electron-hole-plasma is the only origin of the transient intensity reduction detected by the photo diode. The local excess charge concentration *n(y)* at the beam position can be calculated from the absorption law:

$$I_{on}(y) = I_{off}(y) \exp\left(-\partial\alpha / \partial n \cdot L \cdot n(y)\right) \tag{1}$$

The absorption coefficient α depends linearly on the excess charge over a wide range of concentrations (10) and its dependence on the carrier concentration $\partial\alpha/\partial n$ has been calibrated by commutation experiments (11). Scanning the sample yields the excess charge carrier profile. Therefore, the sample is mounted onto a sample holder on a three axis-positioning system in a cryostat. Cooling with liquid nitrogen or helium or using the resistive heater enables the operation of the DUT at temperatures from 40 K to 450 K.

Description of the samples

Two types of freewheeling diodes with platinum doping for lifetime adjustment have been analyzed. First, the anode has been processed on the front side of a float-zone silicon wafer. Platinum has been deposited on the wafer front side and a platinum silicide has been formed by annealing the diode. The residual platinum has been removed by wet etching. The wafer has been thinned to its final thickness by grinding and etching the wafer back side and the cathode has been formed by implantation with a subsequent high-temperature step. Two diode types with a thickness of 650 μm and 110 μm have been considered. The lifetime adjustment has then been carried out by a diffusion step in the temperature range between 750 °C and 850 °C with the platinum silicide at the front side acting as platinum source. Finally, metal contacts have been formed on the wafer front and back side.

For the FCA measurement, small rectangular samples have been cut from the wafer. For the diodes with 650 μm chip thickness the contact area has been 2.5 × 4.0 mm². In this case the spatial resolution is about 20 μm and is limited by the minimum spot diameter on the side facets and depends on the numerical aperture and the light transmission length (2.5 mm). In order to achieve a higher spatial resolution of roughly 10 μm for the diode with only 110 μm chip thickness, this diode has been cut to small peaces of only 1.0 × 2.5 mm². The side facets of the samples have been lapped (grain diameter ≈ 1 μm) and chemical mechanically polished in order to reduce surface recombination effects. Stable oxides were formed on the side facets by an annealing step in air for 10 min at 150 °C.

Experimental results from FCA measurements

A typical carrier concentration profile obtained from the FCA measurements is shown in figure 2 for the 110-μm thick sample. Under forward bias conditions, the diode is operated under high-injection conditions, with the local electron concentration Δn being nearly identical to the local hole concentration Δp ($\Delta n \approx \Delta p$). The current transport in the drift layer is dominated by ambipolar carrier diffusion which is described by the ambipolar diffusion equation:

$$\frac{\partial^2 n}{\partial y^2}(y) = \frac{n(y)}{L_a^2(y)}$$

[2]

This means that the local curvature of the carrier concentration profile $n(y)$ is given by the local ambipolar diffusion length $L_a(y)$, which is related to the ambipolar carrier lifetime τ_a by:

$$\tau_a(y) = L_a^2(y)/D_a$$

[3]

Here, $D_a = 2\dfrac{kT}{q}\dfrac{\mu_n\mu_p}{\mu_n+\mu_p}$ denotes the ambipolar diffusion coefficient with the electron and hole mobility, μ_n and μ_p, Boltzmann's constant k and the elementary charge q. Since the charge carrier concentration is smaller than 10^{17} cm^{-3} (fig. 2) carrier-carrier scattering can be neglected and the mobilities are independent of the carrier concentrations. Supposing in a first approximation a homogeneous carrier lifetime distribution in the drift region, equation [2] can be easily solved, resulting in a cosh-shape of the carrier concentration profile:

$$n(y) = n_{\min} \cosh\left(\frac{y - y_{\min}}{L_a}\right) \qquad [4]$$

Here, $n_{\min}$ denotes the minimum carrier concentration of the profile with its position $y_{\min}$. By fitting this function to the experimental profiles (fig. 2) the diffusion length and the ambipolar carrier lifetime can be extracted. It should be noted that only the relative curvature and not the absolute carrier concentration level is important for the extraction of the diffusion length. Experimental errors in the calibration of the absorption coefficient have no influence on the accuracy of the extracted diffusion length.

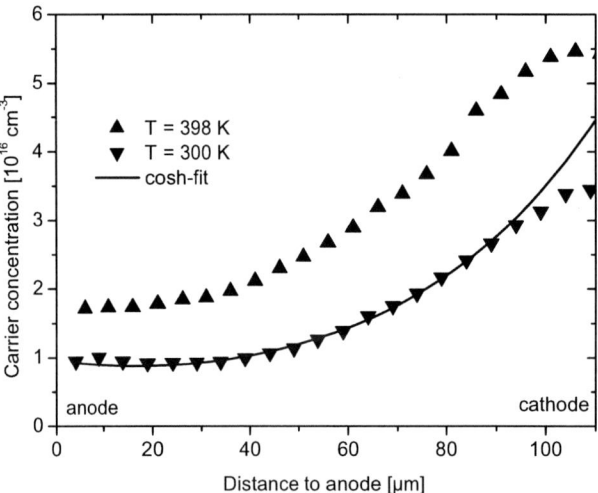

Figure 2. Vertical carrier concentration profiles of the 110-µm thick sample under a forward current pulse of 250 A/cm². Symbols indicate the carrier concentration extracted from FCA measurements. The line represents the fit of a cosh-shaped function to extract the ambipolar diffusion length.

By varying the amplitude of the forward current pulse the injection level in the drift region is modulated. As can bee seen from figure 3, the carrier lifetime is nearly constant over the whole measured range of injection levels. This indicates that Auger recombination does not take place for carrier concentrations smaller than 10^{17} cm^{-3}. The lower limit of the detectable injection level is given by the relative sensitivity of the photo detector and is limited to 10 to 50.

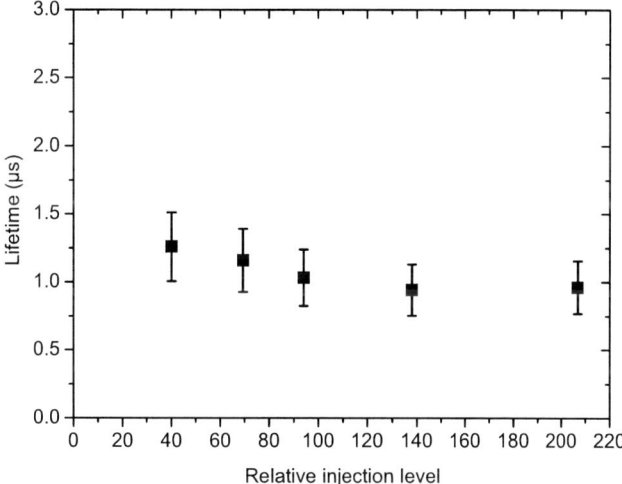

Figure 3. Dependence of the ambipolar lifetime on the injection level at 300 K. The base doping concentration of the sample was 6.6×10^{13} cm^{-3}. The injection level is given as ratio of the minimum excess charge concentration to the base doping concentration.

The dependence of the carrier lifetime on the platinum diffusion temperature has been determined for samples processed with different diffusion temperatures (fig 4.). Obviously, the carrier lifetime decreases with increasing platinum diffusion temperature. The points for the platinum diffusion temperature T_0 to $T_0 + 40$ °C ($T_0 \approx 750$ °C) have been obtained for the 650-μm thick samples. By supposing that the Pt concentration is proportional to the Pt solubility, and assuming temperature-independent capture cross-sections for the relevant Pt acceptor center and a temperature dependence of the thermal velocity $\sim T^{1/2}$, the solution enthalpy $H_S = 2.84$ eV can be extracted from the $\tau(T_{Pt})$ dependence for these points. The calculated value of the solution enthalpy is slightly larger than the enthalpy value of 2.68 eV published in Ref. (9). For the dependence of the carrier lifetime on the operating temperature we have already shown that the results from FCA measurements are consistent to the center parameters obtained by μ-PCD (12).

The point at $T_0 + 66$ °C in figure 4 has been obtained for the sample with 110 μm thickness. The carrier lifetimes of 2 μs at 398 K and 1 μs at 300 K are well below the value which would be expected from the fit at 300 K for the thicker samples. We suppose that this difference indicates that the platinum concentration in the thinner sample is higher than in the 650-μm thick samples, since surface recombination effects and preparation effects can be excluded as possible root cause.

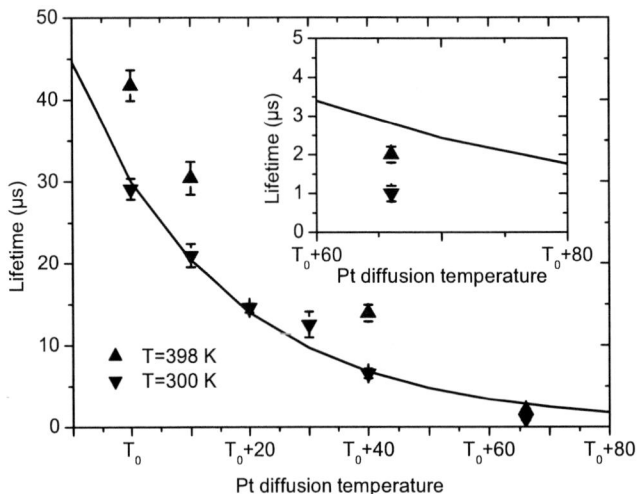

Figure 4. Dependence of the ambipolar carrier lifetime on the platinum diffusion temperature. The symbols indicate the values obtained from FCA measurements at two different operating temperatures. The line indicates a fit for a solid solubility model. In the insert, the axes are rescaled for a better illustration of the results from the 110-μm thick sample.

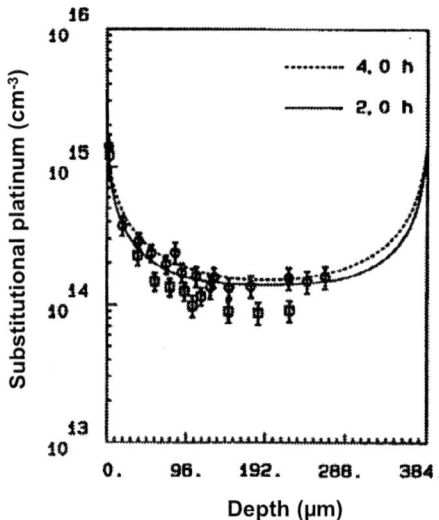

Figure 5. Substitutional platinum profile after ref (13). Circles and squares refer to a sample which has been annealed at 850 °C for 4 h and at 850 °C for 30 min, respectively. The lines correspond to simulations at 850 °C for 2 h and 4 h.

It is well-known that substitutional platinum can exhibit a *U*-shaped profile as shown in fig. 5 [from ref. (13)]. In ref. (13), float-zone silicon samples have been deposited with platinum from a liquid solution. An annealing step at 250 °C for 20 min removed the remaining solvent of the platinum solution. After this annealing step, the platinum diffusion has been carried out in the temperature range between 700 °C and 950 °C for diffusion times ranging from 0.5 h to 142 h. Platinum profiles have been determined by chemically beveling the sample, over-etching the beveled surface, forming Schottky-contacts on the beveled surface and ohmic contacts on the back side, and performing DLTS measurements. From simulations of the point defect diffusion it is concluded that the Frank-Turnbull reaction is the dominant mechanism for platinum activation/diffusion in the temperature range between 700 °C and 850 °C and that the initial vacancy concentration instead of the equilibrium vacancy concentration at the diffusion temperature controls the resulting platinum profile.

In contrast to the process described in ref. (13), platinum diffusion of our samples has been carried out from a platinum silicide. There are strong indications that due to stress – induced during the silicide formation – vacancies are injected from the wafer front side and modify the vacancy profile (14). In fact, the direct extraction of the local curvature by equation [2] and hence the local lifetime profile for our 650-μm thick samples suggest an inverse *U*-profile for the carrier lifetime distribution (12). Here, the carrier lifetime distribution was significantly reduced within the first 50 μm to 100 μm in front of the anode. This evaluation, however, requires the calculation of a spatial second-order derivation and is only reasonably accurate for thicker samples with a sufficient amount of measurement points. The fitting of a cosh-shaped carrier concentration profile however is most sensitive to the nearly constant lifetime in the bulk where the vacancies should reach their equilibrium value. We suppose that for the sample of the 110-μm thick diodes the total platinum profile is governed by the vacancies injected during the platinum silicide formation. Hence, we expect a steep platinum profile decreasing from the anode with a higher concentration. This would explain the shorter carrier lifetime in our thinner sample compared to the extrapolation obtained by the thicker samples (fig. 5).

Determination of platinum profiles by DLTS

To further confirm our hypothesis, we have performed DLTS measurements at the 110-μm thick sample. The depth information has been determined by wet etching the diode from the frontside (anode) or backside (cathode) to various depths and depositing Schottky contacts on the etched surface. Ohmic contacts have been provided on the opposite sample side. CV (Capacitance-Voltage) and DLTS measurements have been performed to determine the doping level and the trap concentration. DLTS spectra obtained at different depths are shown in figure 6. The highest trap concentrations occur at the wafer front side. Peak I can be attributed to the platinum acceptor level at E_c-0.23 eV; this center dominates recombination processes. The energetic level of peak III is E_c-0.55 eV and is close to the well-known hydrogen-related level Pt-H_1 at E_c-0.50 eV as reported in Refs. (12, 15-16).

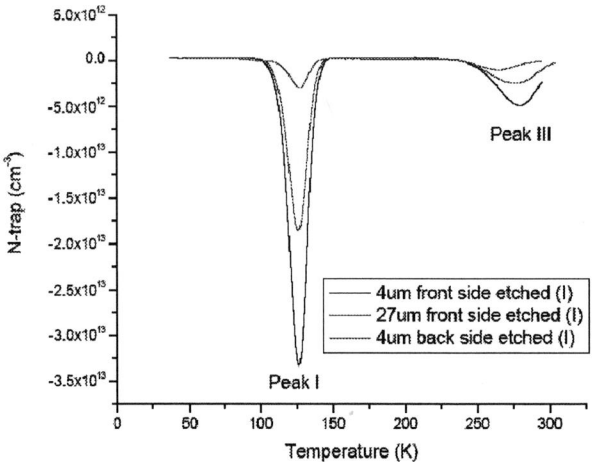

Figure 6. DLTS spectra of a 110-μm thick diode at various depths.

The full trap profile was obtained by successive measurements at different etch depths and is shown in figure 7. The trap concentration of the acceptor level is characterized by a significant decrease from the anode to the cathode side by nearly an order of magnitude. The same is true for the hydrogen-related level Pt-H_1, although its absolute concentration is at least half a decade smaller than that of the Pt-acceptor level. This confirms our hypothesis of a non-uniform platinum profile for the thinner sample.

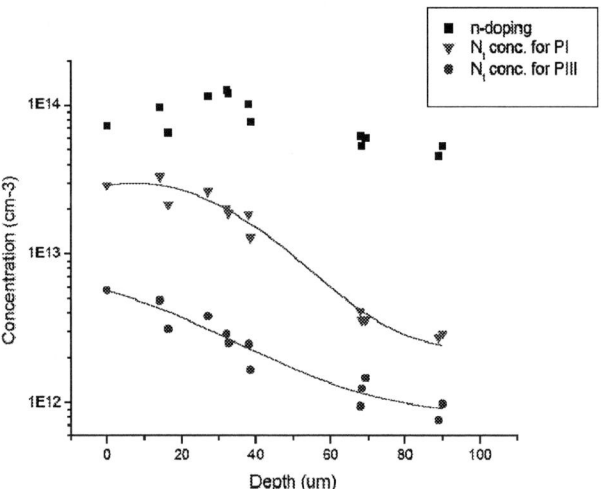

Figure 7. Spatial trap distribution of the 110-μm thick diode. Squares: doping concentration obtained from CV measurements. Triangles: platinum acceptor level. Circles: hydrogen-related level Pt-H_1. Solid lines: Gaussian fit to the measured trap concentration profiles.

Conclusion

We have demonstrated that the free carrier absorption technique is an extremely valuable tool to directly determine the excess charge profile in vertical power devices. The carrier lifetime adjustment in the drift region corresponds to the local curvature of the excess charge carrier profile. This enables the study of the dependence of the carrier lifetime on various parameters. We have shown that the carrier lifetime is independent on the injection level for excess charge concentrations below 10^{17} cm^{-3}. Also, the solution enthalpy for a solid solubility model could be extracted from measurements of 650-μm diodes fabricated with different platinum diffusion temperatures. Our measurements of 110-μm thick samples suggest that the substitutional platinum concentration is larger than it would be expected for thicker samples. We assume that 650-μm thick samples exhibit a U-shaped platinum profile with higher platinum concentration at the anode. For the 110-μm thick diodes, only this first part of the U-shaped platinum profile with increased platinum concentration is supposed to be present. Complementary DLTS measurements on subsequently etched samples allow for the extraction of the deep level profile within the drift layer. These measurements actually confirm our hypothesis of a steeply decreasing platinum profile from the anode for the 110-μm thick sample. The centers found are attributed to the platinum-acceptor level and the platinum-hydrogen level Pt-H$_1$.

Together with our previous work (12) these results provide the basis for self consistent electro-thermal device simulations taking into account the properties of platinum induced deep levels and their spatial distribution.

Acknowledgments

The authors would like to thank L. Palmetshofer (Solid State Physics Division, Johannes Kepler University Linz, Austria) and his staff for performing the spatially resolved DLTS measurements.

References

1. Y. Kirino, A. Buczkowski, Z.J. Radzimski, G.A. Rozgonyi, and F. Shimura, *Appl. Phys. Lett.*, **57**, 2832 (1990).
2. S.H. Gamal, M.L. Locatelli, and J.P. Chante, *EPE Journal*, **2**, 85 (1992).
3. K. Watanabe and C. Munakata, *Semicond. Sci. Technology*, **8**, 230 (1993).
4. K. Watanabe, *Semicond. Sci. Technology*, **11**, 1713 (1996).
5. B. Deng and H. Kuwano, *Jpn. J. Appl. Phys.*, **34**, 4587 (1995).
6. A. Hallen, N. Kesikitalo, F. Masszi, and V. Nágl, *J. Appl. Phys.*, **79**, 3906 (1996).
7. H. Bleichner, P. Jonsson, N. Keskitalo, and E. Nordlander, *J. Appl. Phys.*, **79**, 9142 (1996).
8. N. Keskitalo, P. Jonsson, K. Nordgren, H. Bleichner, and E. Nordlander, *J. Appl. Phys.*, **83**, 4206 (1998).
9. K. Graff, *Metal Impurities in Silicon Device Fabrication*, 2nd rev. ed., Springer, Berlin (2000).
10. H. Bleichner and M. Rosling, *Proc. ISPSD 1990 Tokyo*, p. 246, Tokyo (1990).
11. C. Mehnert and F. Hille, *Technical report 2000/2*, Munich University of Technology, Chair for Physics of Electrotechnology, Munich (2000).
12. H.-J. Schulze, A. Frohnmeyer, F.-J. Niedernostheide, F. Hille, P. Tütto, T. Pavelka and G. Wachutka, *J. Electrochem. Soc.*, **148**, 655 (2001).
13. H. Zimmermann, *Phd-Thesis*, p. 76, University of Erlangen, Erlangen (1991).
14. P. Pichler, *private communications*, Fraunhofer Institut Erlangen, Erlangen (2006).
15. J.-U. Sachse, E.Ö. Sveinbjörnsson, W. Jost, and J. Weber, *Phys. Rev. B*, **55**, 16176 (1997), and *Appl. Phys. Lett.*, **70**, 1584 (1997).
16. J.-U. Sachse, J. Weber, and E.Ö. Sveinbjörnsson, *Phys. Rev. B*, **60**, 1474 (1999).

ECS Transactions, 10 (1) 151-160 (2007)
10.1149/1.2773985 ©The Electrochemical Society

In-Line Characterization of Heterojunction Bipolar Transistor Base Layers by High-Resolution X-Ray Diffraction

N. D. Nguyen[a], R. Loo[a], A. Hikavyy[a], B. Van Daele[a],
P. Ryan[b], M. Wormington[b] and J. Hopkins[b]

[a] IMEC, Kapeldreef 75, B-3001 Leuven, BELGIUM
[b] Bede plc, Belmont Business Park, Durham, DH1 1TW, UK

The suitability of high-resolution X-ray diffraction (HRXRD) as an in-line measurement tool for the characterization of heterojunction bipolar transistor SiGe base layers and Si cap layers was investigated. We showed that despite of polycrystalline Si on the mask material of patterned wafers, HRXRD measurements performed on an array of small windows yield results which are comparable to those that were obtained on a window which is larger than the size of the source beam, regarding the thickness and the Ge content of the SiGe layers. The possibility to extract layer parameters for active device windows of different sizes was therefore demonstrated. The suitability of HRXRD for in-line measurement of the Si cap thickness was also assessed and the sensitivity of this technique for determining the substitutional boron concentration in SiGe was studied. The detection limit in the monitoring of the active dopant concentration was about 2.7×10^{19} cm^{-3}.

Introduction

The efficient control of the properties of the $Si_{1-x}Ge_x$ base layer in heterojunction bipolar transistors (HBT) is a critical requirement in order to optimize the benefit of the $Si/Si_{1-x}Ge_x$ band offset and to achieve high device performance in bipolar complementary metal oxide semiconductor (BiCMOS) technology (1-4). For production applications, layer uniformities in the range of $1 - 2$ % are required and reproducibility from one epitaxial growth process to the other one has to be of the same order of magnitude. Layer thickness and layer composition (Ge content) are generally measured by Rutherford backscattering spectroscopy (RBS), secondary ion mass spectrometry (SIMS) or photoluminescence (PL). Although these techniques are very well developed, they are not suitable as production measurement tools : RBS and SIMS are destructive analysis methods and PL requires time-consuming cryogenic processes which limit factory throughput. Alternatively, high-resolution X-ray diffraction (HRXRD) and spectroscopic ellipsometry (SE) allow a fast, non-destructive analysis and are therefore suitable for in-line monitoring. Indeed, it has been demonstrated that SE can be used as an in-situ or ex-situ technique to determine the composition and thickness of either thick, relaxed or thin, strained $Si_{1-x}Ge_x$ epitaxial layers while HRXRD is an established technique that can accurately measure strain, composition, relaxation and thickness of $Si_{1-x}Ge_x$ layers grown by epitaxial processes (5, 6).

In this work, we assessed the applicability of HRXRD as a production-oriented measurement tool for the characterization of BiCMOS heterostructures. The study was

151

carried out on both blanket and patterned wafers. It focused on three different applications. In the first application, a wafer was processed with a pattern containing test matrices, with each test matrix consisting of an array of a single window dimension. A SIMS pad, which mimics blanket-like measurement conditions, was included in the pattern. As we report here, although polycrystalline Si is present on the mask material, the interpretation of the ω-2θ scans performed on an array of small windows leads to similar conclusions as those that are obtained on a window which is larger than the size of the incident X-ray beam. From the point of view of device applications, this means that HRXRD allows to control the epilayer properties for different device sizes.

It is also important that the values of the layer thicknesses are correctly extracted from the analysis of the HRXRD scans. In order to assess the suitability of HRXRD for in-line measurement of the Si cap thickness in HBT base layers, comparison with SE was done on Si/SiGe layer stacks with different deposition times for the growth of the Si-cap. In the third application, the sensitivity of HRXRD for the determination of the substitutional boron concentration in SiGe was investigated. The distortion of the silicon lattice due to doping with boron as a substitutional impurity has been studied since the early days of silicon technology, leading to highly accurate measurements of the lattice contraction coefficient (7-11). In this paper, our aim is to give an estimation of the minimum active boron concentration in SiGe that can be monitored by HRXRD in production environment. For this purpose, measurements of the XRD peak shift for layers with boron content varying within a broad range are needed. Smooth epilayers were fabricated by in-situ doping of SiGe grown by reduced pressure chemical vapor deposition (RPCVD) which resulted in highly homogeneous boron concentrations. The high quality of the epitaxially-deposited films allowed to avoid the difficulties that could be encountered in the interpretation of XRD data of diffused materials with non uniform boron contents (11), and to determine the concentration threshold for HRXRD monitoring with good reliability. Finally, this utility of HRXRD can be exploited in other applications like embedded SiGe which is nowadays used in pMOS devices to boost device performance (12).

The paper is organized as follows. After a description of the growth and characterization procedures, we first present the results obtained by HRXRD on patterned wafers and discuss the comparison between small and large windows as well as comparison with SIMS data. Next the correlation between thickness measurements as extracted from HRXRD and those extracted from SE is discussed. Finally, we demonstrate the ability of HRXRD to monitor the active boron concentration and give an estimation of the threshold value.

Experimental

All epitaxial layers were deposited on 200 mm Si (001) wafers, using a standard horizontal cold wall, load-locked, ASM Epsilon™ 2000 reactor, a RPCVD system designed for production applications. Deposition conditions for non-selective epitaxial growth of Si and $Si_{1-x}Ge_x$ include a pressure of 40 Torr and H_2 as a carrier gas. Silane (SiH_4) and germane (GeH_4, 1 % diluted in H_2) were used as Si and Ge source gases, respectively, while diborane (B_2H_6, 50 ppm or 1 % diluted in H_2) was used as source gas for boron (B). For patterned wafers, a combination of 20 nm oxide capped with 60 nm nitride was utilized as mask material. The test matrices of the pattern have dimensions of

270 μm × 270 μm whereas the dimensions of the open windows varied from 0.18 μm to 10 μm in both directions. Before deposition, the blanket wafers received a NH_4OH/O_3-based clean followed by an in-situ bake at 1050°C for 60 seconds in H_2 in order to remove the native oxide. For patterned wafers, the pre-deposition treatment consisted in a NH_4OH/O_3-based clean followed by a HF 2% dip during 30 seconds and an in-situ bake at 850°C for 2 minutes. Because of the non-selective conditions of the growth, deposition occurs simultaneously in the open Si windows (epitaxial growth) and on the mask material (growth of polycrystalline layers) (13).

HRXRD measurements were performed using a BedeMetrix™-L tool from Bede X-ray Metrology fitted with a Microsource™ micro-focus source, a ScribeView™ optic and a channel-cut beam conditioning crystal. The beam cross-section at the sample position was less than 100 μm × 100 μm. The angular position ω of the sample' axis and that of the detector axis, 2θ, were adjusted with respect to the incident X-ray beam using a high-precision goniometer (6). The background intensity was less than 0.3 counts per second (cps). The analysis of the HRXRD spectra was performed using the RADS 4.0 software, a simulation and data-fitting program from Bede X-ray Metrology (14, 15). SE measurements were made with an Advanced Spectroscopic Ellipsometry Technology (ASET-F5) system from KLA-TENCOR, which is a production-oriented completely automated small-spot (30 μm × 30 μm) spectroscopic ellipsometer. The polarizer rotates continuously and the analyzer is fixed in position of each measurement. The SE technique consists of measuring the (tan Ψ, cos Δ) spectra, which are collected as function of wavelength in the range 250 – 750 nm from the integrated intensity reaching the detector in each 45° octant of the polarizer's rotation, and performing a mathematical regression analysis using the harmonic oscillator model developed by C. Ygartua and M. Liaw (16). Within this method, single $Si_{1-x}Ge_x$ epilayers and $Si/Si_{1-x}Ge_x$ layer stacks can be studied and layer thicknesses as well as Ge contents can be determined (5). By comparison with stepheight measurements, the accuracy of the determined thicknesses is about 2 %. The SIMS was performed on an Atomika 4500 depth profiler. Measurement conditions consisted of 500 eV O_2^+ bombardment at 0° incidence.

Results and Discussion

First we discuss the results obtained on patterned wafers, the open windows of which have been filled with an epitaxially-deposited full HBT base layer. HRXRD data measured from an array of small windows are compared to those measured from a large pad and to SIMS results. The conclusions of this work will give the motivation for the next experiment which consists in comparing the values of the Si-cap thickness measured by HRXRD, SE and SIMS. In the last part, we will discuss the monitoring of the boron concentration by means of HRXRD and determine the threshold concentration for the active boron.

<u>Determination of Layer Thickness and Ge Content of HBT Layers Deposited on Patterned and Unpatterned Wafers with Non Selective Growth Conditions</u>

In the open windows, the typical full base layer stack for HBT in BiCMOS applications contains a first epitaxial $Si_{1-x}Ge_x$ layer (labeled L1) deposited on the Si substrate, followed by a second $Si_{1-x'}Ge_{x'}$ layer (labeled L2) with x' < x. A boron spike doping is included in the first $Si_{1-x}Ge_x$ layer. The stack is capped by a pure Si layer (labeled L3). On the mask material, the RPCVD process conditions resulted in the growth of polycrystalline material. The compositional analysis was done on a large (270 μm × 270 μm) open area of the pattern (SIMS pad). The Si, Ge and B concentrations as function of depth are shown in Figure 1a.

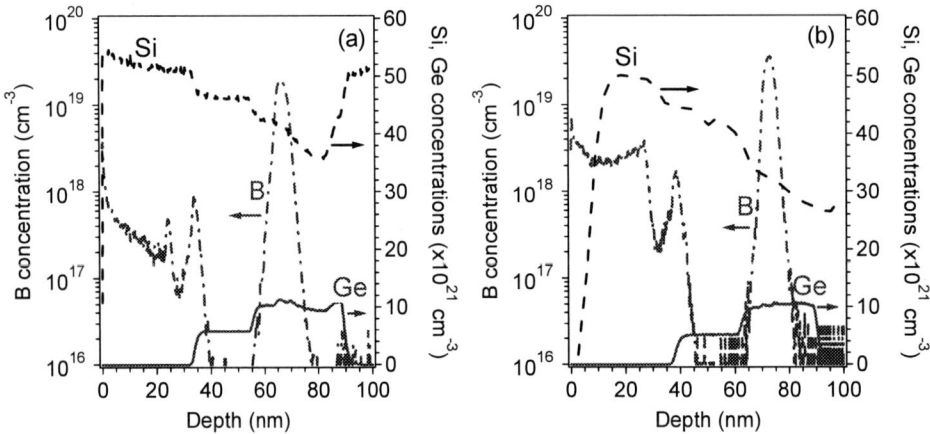

Figure 1. Concentrations of Si, Ge and B in the base layer of a typical npn HBT as function of depth, as measured by SIMS, (a) in the SIMS pad of a patterned wafer and (b) on a blanket wafer.

As expected from the definition of the deposition sequence, two steps are present in the Ge curve. Since the goal is to check the correlation between different characterization techniques for the determination layer thickness and layer composition, the design of the base in this work does not include any ramped Ge layer although graded Ge alloying in the base allows to reduce the transit time by forming an accelerating field for the minority carriers (17). In Figure 2a, the anomalies in Si and Ge profiles are due to charging effects and do not have any impact on the thickness values. The boron profile shows a main peak in layer L1 at 66.5 nm from the surface with a magnitude of 2×10^{19} cm^{-3} and a full width at half magnitude (FWHM) of 4.8 nm. For the determination of the thicknesses, we applied the following cutoff condition, which is widely used, to the Ge profile : the boundary between two adjacent layers is defined as the depth coordinate at which the Ge concentration is the average of their values. Prior to this, the nominal Ge concentration in layers L1 and L2 were calculated by averaging the data points of each plateau. The results of the analysis are shown in Table I. The typical error relative to thickness estimation from given SIMS profiles is about 0.2 nm.

Figure 2 shows typical 004 ω-2θ scans from an array of 5 μm × 5 μm wide windows (Figure 2a) and from a 270 μm × 270 μm SIMS pad (Figure 2b). The latter one corresponds to a virtual measurement on blanket wafer. The scans exhibit two main different diffraction peaks; the most intense and narrowest peak corresponds to the Si substrate while the broader structure containing interference fringes can be attributed to the SiGe layers. The HRXRD data were analyzed using dynamical diffraction theory and an automated data-fitting algorithm embedded in RADS software from Bede X-ray Metrology (14, 15). Agreement between experiment and best-fit simulation of the $Si/Si_{1-x'}Ge_{x'}/Si_{1-x}Ge_x$ epitaxial structure is very good.

TABLE I. Thickness and Ge content of the HBT base layer						
		Patterned wafer			**Blanket wafer**	
		HRXRD on 5 μm × 5 μm	HRXRD on SIMS pad	SIMS	HRXRD	SIMS
Thickness (nm)	L3	39.6 ± 0.4	41.2 ± 0.2	34.4	37.2 ± 0.4	38.0
	L2	21.8 ± 0.3	23.4 ± 0.2	22.0	22.3 ± 0.3	25.6
	L1	33.8 ± 0.3	36.3 ± 0.1	34.0	27.2 ± 0.3	26.8
Ge (%)	L2	6.26 ± 0.12	6.45 ± 0.06		7.11 ± 0.06	
	L1	13.22 ± 0.03	13.58 ± 0.02		12.92 ± 0.03	

As shown in Table I, the results obtained for the array of small windows are very similar to those corresponding to the SIMS pad. It must be highlighted here that this correlation was observed despite of the polycrystalline Si present on the mask in the first case. Figure 2 also illustrates that the presence of the non-monocrystalline material does not affect the angular position of the peaks nor the quality of the fit; it only alters the relative intensities. The decrease in the coherently diffracted intensity might lead, in some cases, to the reduction of a peak to a shoulder that cannot be resolved in position. Interference fringes are indeed more visible in the experimental data related to the SIMS pad because a weaker effect of the polycrystalline material is expected. Moreover, concerning layer thicknesses, the fact that the converged values are systematically lower for the small windows measurements than for the large window case are also explained by the general intensity reduction of the broader peaks and interference fringes. Despite the abovementionned remarks, a good overall agreement of the values of the thicknesses and of the Ge contents could be reached between results obtained for the array of 5 μm × 5 μm windows and those obtained for the large window. From the point of view of device applications, this means that HRXRD can be efficiently used as an in-line characterization technique to monitor epitaxial layer stacks. An efficient control of the epilayer properties for different device sizes is therefore possible with this method.

The comparison with the SIMS results for the SiGe thicknesses shows a good correlation with HRXRD data. This confirms the satisfactory quality of the latter measurements. A significant disagreement between HRXRD and SIMS can however be observed for the Si-cap thickness. The average difference is about 6.0 nm. Such a discrepancy can not be attributed alone to surface transient effects which are known to alter secondary ion yields (18-22). Even if we take into account the full error corresponding to the accuracy of thickness determination from the concentration profiles and add it to the SIMS differential shift, the observed deviation is still too large. This is in contrast with the good correlation that was observed in a similar experiment on blanket wafer. The values of the layer thicknesses and Ge contents for a full HBT layer stack grown on an unpatterned wafer are given in Table I. The concentrations as functions of depth are shown in Figure 1b; here, the measurement of the Si profile was performed with a lower resolution but layer thicknesses can still be determined by the analysis of the Ge profile. The value of the Si-cap thickness as extracted from HRXRD data and that as extracted from SIMS are particularly in very good agreement for the blanket wafer whereas they are significantly different for the patterned wafer. In the latter case, the comparison is done for HRXRD measurement on a large window. In order to clarify this issue, we designed an experiment to specifically assess the quality of the Si-cap thickness measurement by HRXRD.

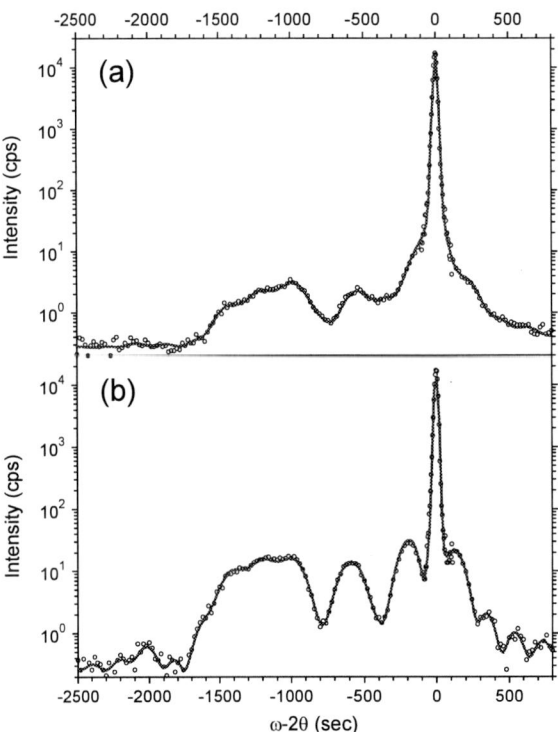

Figure 2. HRXRD scans from (a) a 5 μm × 5 μm test pad and (b) a 270 μm × 270 μm SIMS test pad. The measured ω-2θ scans (circles) around the symmetric 004 reflection are shown with their best-fit simulations (full lines).

Determination of the Si-cap Thickness : Comparison Between HRXRD, SE and SIMS

First a single SiGe reference layer was epitaxially grown on p-type Si substrate. Indeed, for our fitting procedure, the layer stack must be restricted to a single-step Ge profile in order to enable SE measurements. Then we fabricated Si/SiGe layer stacks with various Si deposition times. All other growth conditions were identical, including those used for the SiGe layer. Deposition was done on blanket material. From SE characterization of the reference wafer, we obtained an epilayer thickness at the center point of 98.5 nm and a Ge content of 16.2 %. A 49-point line scan measurement gave an average value of 99.7 nm and 16.0 % of Ge. From HRXRD data collected at the center point of the wafer, a Ge content of 16.3 % was determined whereas best-fit simulation yielded a thickness value of 96.3 nm and 16.4 % of Ge. The standard deviation of the layer thickness as measured by the SE line scan was 1.3 %.

The results of the Si-cap thickness measurements are given in Table II. A first batch consisting of 4 wafers (#1 - #4) was fabricated. The SE average values and standard deviation (std dev.) illustrate the good uniformity of the epitaxial deposition whereas single point comparison between HRXRD and SE were done at the center of the wafer. The results show that both techniques are in good agreement regarding the determination of the Si cap thickness. The discrepancy ranges from 0.8 nm for the thickest (~ 110 nm) Si epilayer to 1.1 nm for the thinnest one (~ 22 nm).

In order to confirm the quality of that correlation, a second batch of 3 wafers (#5 - #7) was processed. A new SiGe reference wafer was used for this run. From single point SE measurement we obtained a value of 89.7 nm for the SiGe thickness and Ge content of 16.3 % at the center of the wafer and an average of 89.2 nm with 16.2 % of Ge was deduced from a 49-point line scan. The SiGe layer was uniform within standard deviation of 0.8 %. The measurements of the Si cap thicknesses of the second batch by the two techniques are included in Table II. Here, the discrepancy ranges from 0.8 nm for the thickest layer to 1.2 nm for the thinnest one.

TABLE II. Si cap layer thickness (nm)

Wafer	HRXRD (center)	SE (center)	SIMS	SE (average)	SE std dev. (%)
#1	21.6	22.7		21.5	4.5
#2	26.3	26.2		24.3	6.8
#3	55.1	55.0		54.0	2.3
#4	110.0	109.2		112.9	2.4
#5	18.5	19.7	18.6	18.6	4.1
#6	46.8	46.4	44.7	44.0	3.5
#7	91.8	91.0	88.4	88.5	1.9

Figure 3 shows a graphical summary of the results. It illustrates the excellent correlation between HRXRD and SE for the evaluation of the thickness of the Si cap layer. These experiments clearly demonstrate that the determination of the Si cap thickness by HRXRD is as accurate and reliable as by SE, which has had its reliability demonstrated in literature for the characterization of single SiGe epilayers and Si/SiGe layer stacks (5, 16). SIMS analysis was also performed on wafers #5 to #7 and the cap layer thickness was estimated from the Ge concentration profiles (not shown here); results are given in Table II.

The comparison of HRXRD, SIMS and SE data leads to the observation that overall, there is a good agreement between the three techniques for measurements on blanket materials. For patterned wafers, the discrepancy between HRXRD and SIMS measurements could be explained by a couple of reasons. At the Bragg angle for Si (004), the footprint of the beam is elliptical, the size being doubled in one direction. This can lead to an extension of the incident intensity outside of the SIMS pad and to a decreasing of the signal-on-noise ratio. Furthermore, edge effects might reduce the effective uniform area of the SIMS pad and contribute to the increasing of the background.

Monitoring of the Active Boron Concentration in SiGe by HRXRD

The incorporation of boron at high concentration levels causes a significant lattice distortion of the doped layer with respect to the undoped material, resulting from the smaller covalent radius of boron compared to that of silicon or germanium. The lattice mismatch induces strain and can generate misfit dislocations as well in thick, relaxed layers. In the regime of purely elastic accommodation of the lattice to the incorporated boron, the layer strain depends linearly on the dopant concentration (23, 24). Since a linear response is desired for any sensitive monitoring, measurement of the peak shift by HRXRD, which is directly related to the strain in the epilayer, proved to be an efficient method to determine the active boron concentration. For in-line characterization, it is necessary to know the concentration range that can be monitored.

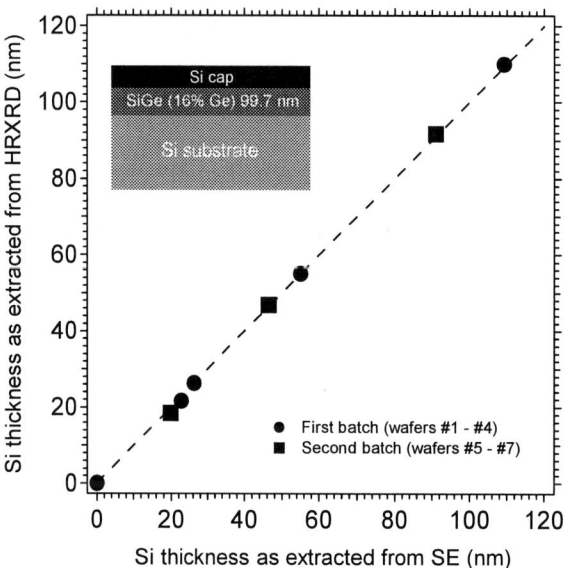

Figure 3. Measurements of the Si cap thickness : comparison between results extracted from HRXRD and those from SE. The layer stack of the first batch is shown as inset. A similar layout was used for the second batch, except for the properties of the SiGe buried layer. The dashed line represents the case of a perfect match between HRXRD and SE.

Therefore, as final application, the sensitivity of HRXRD for monitoring the substitutional boron concentration in SiGe was investigated. An undoped SiGe was first grown. SE measurement performed at the center of the wafer provided a layer thickness of 98.9 nm and Ge content of 16.2 %. From a 49-point line scan, an average thickness of 100.0 nm was measured with a standard deviation of 1.2 % and 16.1 % of average Ge. The best-fit simulation of the HRXRD data provided a thickness value of 96.7 nm and Ge content of 16.4 %. For doped SiGe layers, the control of the active dopant concentration was performed by the flow of the diborane gas. All depositions were made on n-type blanket Si wafers. The substitutional boron concentration for each SiGe:B layer was extracted from the value of the sheet resistance as measured by the four-point probe (4pp) method and by using the average thickness provided by SE. SE and HRXRD were in good agreement regarding the measurement of the undoped and doped SiGe layer thicknesses (not shown here). The Ge content was practically constant for all doping conditions whereas the hole concentration increased with the diborane flow following a power law for moderate concentration values but showing saturation at about 1.6×10^{20} cm^{-3} obtained for the highest gas flows. This is an indication of the solid solubility of boron in SiGe with Ge content of about 16 %.

The result of the assessment is shown in Figure 4. A linear dependence of the XRD epilayer peak position on the active boron concentration was found for high boron content whereas for low concentrations, the position of the peak is constant. From the intersection of the linear fit and the averaging, the threshold value is estimated at 2.7×10^{19} cm^{-3}. From SIMS analysis, we deduce that this corresponds to an activation level of about 50 %. The measured concentration profiles allowed to check the very good uniformity of the layer doping. At even higher boron concentrations, the linear trend is not expected to continue as boron would start to precipitate out of solid solution leading to a reduction in the strain gradient with B concentration.

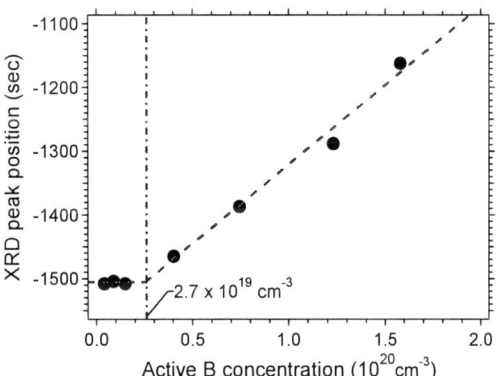

Figure 4. Position of the XRD peak as function of the active boron concentration. Experimental data points are shown in full circles. Dashed lines represent the linear fits and the intersection between the two regions is shown by the dash-dotted line, which corresponds to a threshold concentration of about 2.7×10^{19} cm^{-3}.

Conclusions

In this work, we discussed results of the characterization of HBT base layers by HRXRD and compared the results obtained with those from SE and SIMS measurements. The goal was to assess the suitability of HRXRD as in-line measurement method to monitor Si and SiGe layer thickness as well as layer composition (Ge and B content) in production environment. Specifically for that purpose, HRXRD benefits from several advantages over established techniques. One of them is that it allows for the analysis of the complete HBT layer stack whereas SE is restricted to single SiGe layers or Si/SiGe layer stacks. It is a fast and non-destructive analysis tool that can be readily used to measure small windows on real devices whereas SIMS is limited by a much lower throughput and the device is lost after measurement. We showed that HRXRD measurements of the thickness and Ge content of SiGe layers performed on an array of small windows yield results which are comparable to those that can be obtained on a large window, despite of the presence of polycrystalline Si on the mask material. We also showed that the accuracy of this technique is as good as that of SE concerning the determination of the Si-cap layer thickness on blanket wafers. For patterned wafers, HRXRD measurements led to an overestimation of 6 nm. Finally, the ability of HRXRD to monitor the active boron concentration in SiGe layers was demonstrated for boron contents higher than 2.7×10^{19} cm^{-3}.

Acknowledgments

This work was achieved within the framework of a Joint Development Program between Bede X-ray Metrology and IMEC.

References

1. A. Gruhle, in Proceedings of the 2001 Bipolar/BiCMOS Circuits and Technology Meeting (BCTM), IEEE, p. 19 (2001).
2. R. Loo, M. Caymax, I. Peytier, S. Decoutere, N. Collaert, P. Verheyen, W. Vandervorst, and K. De Meyer, *J. Electrochem. Soc.*, **150**, G638 (2003).
3. P. Wennekers and R. Reuter, in J. D. Cressler, in *Proceedings of the 2004 Bipolar/BiCMOS Circuits and Technology Meeting (BCTM)*, IEEE, p. 79 (2004).
4. J. D. Cressler, in *Proceedings of the 2005 Bipolar/BiCMOS Circuits and Technology Meeting (BCTM)*, IEEE, p. 248 (2005).
5. R. Loo, M. Caymax, G. Blavier, and S. Kremer, *Proc. SPIE 4406*, edited by G. Kissinger and L. H. Weiland, p. 131 (2001).
6. M. Wormington, T. Lafford, S. Godny, P. Ryan, R. Loo, A. Hikavyy, N. Bhouri, and M. Caymax, submitted to Frontiers of Nanoelectronics 2007.
7. G. L. Pearson and J. Bardeen, *Phys. Rev.*, **75**, 865 (1949).
8. F. H. Horn, *Phys. Rev.*, **97**, 1521 (1955).
9. T. Fukumori, K. Futugami, and K. Matsunaga, *Jpn. J. Appl. Phys.*, **21**, 1525 (1982).
10. H.-J. Herzog, L. Csepregi, and H. Seidel, *J. Electrochem. Soc.*, **131**, 2969 (1984).
11. H. Holloway and S. L. McCarthy, *J. Appl. Phys.*, **73**, 103 (1993).
12. T. Ghani, M. Armstrong, C. Auth, M. Bost, P. Charvat, G. Glass, T. Hoffmann, K. Johnson, C. Kenyon, J. Klaus, B. McIntyre, K. Mistry, A. Murthy, J. Sandford, M. Silberstein, S. Sivakumar, P. Smith, K. Zawadzki, S. Thompson, and B. Bohr, *IEDM Techn. Dig.*, 978 (2003).
13. W. B. de Boer and D. Terpstra, in *Advances in Rapid Thermal Processing*, F. Roozeboom, J. C. Gelpey, M. C. Öztürk, and J. Nakos, Editors, PV 99-10, p. 309, The Electrochemical Society Proceedings Series, Pennington, NJ (1999).
14. D. K. Bowen, L. Loxley, B. K. Tanner, M. L. Cooke, and M. A. Capano, *Mater. Res. Soc. Symp. Proc.*, **208**, 113 (1991).
15. M. Wormington, C. Panaccione, K. M. Matney, and D. K. Bowen, *Phil. Trans. R. Soc. Lond. A*, **357**, 2827 (1999).
16. C. Ygartua and M. Liaw, *Thin Solid Films*, **313-314**, 237 (1998).
17. G. L. Patton, D. L. Harame, J. M. C. Stork, B. S. Meyerson, G. J. Scilla, and E. Ganin, *Electron Device Lett.*, **10**, 534 (1989).
18. Z. X. Jiang, P. F. A. Alkemade, *Surf. and Interf. Analysis*, **27**, 125 (1999).
19. C. W. Magee, G. R. Mount, S. P. Smith, B. Herner, and H.-J. Gossmann, *J. Vac. Sci. and Technol. B*, **16**, 3099 (1998).
20. J. B. Clegg, *Surf. and Interf. Analysis*, **10**, 332 (1986).
21. K. Wittmaack and I. W. Drummond, *Phil. Trans. R. Soc. Lond. A*, **354**, 2731 (1996).
22. W. Vandervorst and F. R. Shepherd, *J. Vac. Sci. Technol. A*, **5**, 313 (1987).
23. M .Wormington, private communication (2007).
24. B. G. Cohen, *Solid State Electron.*, **10**, 33 (1967).

Author Index

Abbadie, A.	3	Pfeffer, M.	35
Armigliato, A.	57	Pfitzner, L.	35
		Polignano, M.	85
Balboni, R.	57	Pollakowski, B.	51
Bauer, A.	117	Possner, D.	21
Beckhoff, B.	51		
Borionetti, G.	75	Rinaldi, A. M.	75
Bresolin, C.	85	Rink, I.	109
Brunier, F.	3	Rommel, M.	117
		Rucki, A.	97
Cerva, H.	21, 97	Ryan, P.	151
Codegoni, D.	85	Ryssel, H.	117
Corradi, A.	75		
		Schulze, H.	141
Don, E.	35	Schwalke, U.	129
Dontas, I.	65	Soncini, V.	85
Emons, C.	109	Takami, K.	75
Fliegauf, R.	51	Ulm, G.	51
Hartmann, J.	3	Van Daele, B.	151
Hikavyy, A.	151		
Hille, F.	141	Weser, J.	51
Hopkins, J.	151	Wormington, M.	151
Hurlebaus, M.	35	Wyon, C.	35
Kennou, S.	65		
Kluppel, V.	21		
Kolbe, M.	51		
Kolbesen, B. O.	21		
Ladas, S.	65		
Loo, R.	151		
Mainardi, N.	75		
Müller, M.	51		
Nguyen, N.	151		
Niedernostheide, F.	141		
Nutsch, A.	35		
Oechsner, R.	35		
Papaefthimiou, V.	65		
Parisini, A.	57		

The Electrochemical Society, Inc.
65 South Main Street
Pennington, New Jersey 08534-2839, USA

ISBN 978-1-60423-825-9